CITIES ON THE BEACH

Management Issues of Developed Coastal Barriers

Edited by

Rutherford H. Platt
Sheila G. Pelczarski
Barbara K. R. Burbank

*Land and Water Policy Center
Department of Geology and Geography
University of Massachusetts at Amherst*

THE UNIVERSITY OF CHICAGO
DEPARTMENT OF GEOGRAPHY
RESEARCH PAPER NO. 224
1987

Copyright 1987

Rutherford H. Platt, Sheila G. Pelczarski, and Barbara K.R. Burbank

Published 1987 by The Department of Geography

The University of Chicago, Chicago, Illinois

Library of Congress Cataloging-in-Publication Data

Cities on the beach.

    (Research paper / University of Chicago. Dept. of Geography ; no. 224)
    Bibliography: p. 317
    1. Coastal zone management--United States--Congresses.
2. Shore protection--United States--Congresses. I. Platt, Rutherford H. II. Pelczarski, Sheila G. III. Burbank, Barbara K. R. IV. Series: Research paper (University of Chicago. Dept. of Geography) ; no. 224.
H31.C514 no. 224   910 s [333.91'7'09733] 86-25051
[HT392]
ISBN 0-89065-128-0 (pbk.)

Research Papers are available from:

The University of Chicago
The Department of Geography
5828 S. University Avenue
Chicago, Illinois 60637-1583
Price: $10.00; $7.50 series subscription

DEDICATION

The editors wish to express our utmost gratitude to the fourth member of our team, Kim Magnell. Kim as office manager and editorial associate has provided unstinting effort to make the "Cities on the Beach" conference and this volume successful. Her friendly, calm, and professional handling of all sorts of tasks facilitated the entire project. We affectionately dedicate this volume to Kim.

Rutherford H. Platt
Sheila G. Pelczarski
Barbara K. R. Burbank

Table of Contents

INTRODUCTION

Overview of Developed Coastal Barriers
R. H. PLATT 1

THE COASTAL BARRIER RESOURCES ACT

The 1982 Coastal Barrier Resources Act: A New Federal Policy Tack
D. R. GODSCHALK 17

GEOGRAPHICAL CHARACTERISTICS OF COASTAL BARRIERS

A Management-Oriented, Regional Classification of Developed Coastal Barriers
J. K. MITCHELL 31

The Political Geography of Developed Coastal Barriers
R. H. PLATT and K. CALLAHAN 43

Population Changes in Coastal Jurisdictions with Barrier Beaches: 1960-1980
N. WEST 55

Shoreline Changes on Developed Coastal Barriers
K. F. NORDSTROM 65

PLANNING AND GROWTH MANAGEMENT

Managing Change on Developed Coastal Barriers
D. J. BROWER and T. BEATLEY 83

Coastal Barrier Protection and Management: Collier County, Florida
M. A. BENEDICT 95

Regional Multiple Use, Local Single Use: A Potential Model for Regional Coastal Barrier Management
H. NEUHAUSER 107

Management of Adjacent Lands: Lessons from the National Parks
M. MANTELL and C. J. DUERKSEN   113

Campground Towns of the South Carolina Grand Strand
R. L. JANISKEE and P. LOVINGOOD   121

Estimating the Economic Benefits of Beach Recreation
J. SILBERMAN   131

SHORELINE MANAGEMENT

Approaches to Coastal Hazard Analysis: Ocean City, Maryland
S. P. LEATHERMAN   143

Dune Management: Planning for Change
N. P. PSUTY   155

A Successful Local Program for Preserving and Maintaining Dunes on a Developed Barrier Island: Mantoloking, N.J.
P. J. GODFREY   163

Dune Management Recommendations for Developed Coastlines
P. A. GARES   171

Individual Attitudes toward Erosion Policies: Carolina Beach, N.C.
O. J. FURUSETH and S. M. IVES   185

HAZARD MANAGEMENT

Deciding Whether to Evacuate a Beach Community during a Hurricane Threat
J. BAKER   199

Structural Evaluation of Hurricane Shelters
C. P. JONES and B. SPANGLER   211

Hurricane Alicia and the Galveston Experience
J. M. McCLOY and S. N. HUFFMAN   221

Reducing the Psychosocial Trauma of a Hurricane
A. P. CHESNEY   229

LEGAL ISSUES

Constitutional Issues in Post-Hurricane Reconstruction Planning
R. HAMANN   237

The NFIP and Developed Coastal Barriers
    A. D. DAWSON   245

Financing Coastal Barrier Infrastructure
    H. C. MILLER   261

MANAGEMENT ALTERNATIVES

A Time for Retreat
    O. H. PILKEY   275

An Acquisition Program for Storm-Damaged Properties
    on Coastal Barriers: The State Role
    G. R. CLAYTON   281

The Retreat Alternative in the Real World:
    The Kill Devil Hills Land-Use Plan of 1980
    R. P. STURZA II   289

Transferable Land Rights on Developed Coastal Barriers
    J. T. B. TRIPP   295

APPENDICES

CBRA   301

Biographical Sketches   309

Selected Bibliography on Coastal Barriers   317

INTRODUCTION

CITIES ON THE BEACH: AN OVERVIEW

Rutherford H. Platt

University of Massachusetts

Introduction

On January 15-18, 1985, some 250 researchers, government officials, citizen activists, and students gathered at the Pavilion Tower Hotel in Virginia Beach, Virginia to attend a conference entitled: "Cities on the Beach: Management of Developed Coastal Barriers." The conference was organized by the Land and Water Policy Center of the University of Massachusetts at Amherst with funding provided by the National Science Foundation, the Federal Emergency Management Agency, and the National Park Service. The conference opened with a welcome from Governor Charles Robb of Virginia amid a tangle of television crews and newspaper reporters. Then it got down to work in a carefully planned program of plenary and concurrent paper sessions, discussion groups, meals, and a field tour of the local development scene in Virginia Beach (whose population grew by 52% to 262,000 between 1970 and 1980).

Altogether, some 75 speakers presented or discussed papers at the conference. This volume comprises not a verbatim proceedings, but a refined and edited selection of 27 of these papers. The editors regret that space did not permit the inclusion of several more very good papers from the conference. We are indebted to Richard Goulet of the National Science Foundation for orchestrating the funding of the conference (Grant no. OIR- 8312579) and to Bernard Lalor and Katharine M. Walsh of the University of Chicago Department of Geography Research Series for assisting in the publication of this volume.

What is a Coastal Barrier?

Coastal barriers (known earlier as barrier beaches) are the subject of much geomorphological literature (e.g., Kaufman and Pilkey, 1983; Dolan, Hayden, and

Lins, 1980; Godfrey, 1976; Schwartz, 1982). This volume addresses coastal geology only in the context of managing human communities on barriers, as in the papers by Nordstrom, Psuty, Pilkey, Gares, and Leatherman. The following is merely a thumbnail sketch of the physical nature of coastal barriers. The copious writings of the researchers just mentioned should be consulted for further detail.

Coastal barriers fringe low-lying coastal plains along many of the world's oceanic shorelines. One of the most prominent and nearly continuous series of barriers borders the Atlantic and Gulf coasts of the United States, extending about 2,700 miles from southern Maine to Texas. This system consists of about 400 distinct barrier beaches, some of which are true barrier islands and the rest spits and tombolos which are connected to the mainland (U.S. DOI, 1982, p. 4). In physical terms, islands, spits, and tombolos are three forms of the same phenomenon, distinguishable only by the whim of the ocean in moving sand and establishing inlets. All will be subsumed in the term "coastal barrier" for purposes of this volume.

Coastal barriers under natural conditions are characterized by Kaufman and Pilkey (1983, p. 15) as existing in a state of "dynamic equilibrium." This equilibrium represents a balance among four elements of barrier evolution: (1) beach material, (2) energy, (3) sea level, and (4) shape of beach. Beach material consists of sand from rock, shell, or coral origins, depending on location and type of material washed ashore. Energy is supplied through wave, wind, current, and tidal action. Storms and hurricanes subject barriers to transformation through the discharge of high levels of energy. Gradual sea level rise, estimated at about one foot per century along the northeast coast of the United States (Barth and Titus, 1984), causes gradual shoreline erosion, accompanied by narrowing and steepening of barrier beaches. This in turn exacerbates the short-term impact of ocean storms, increasing the extent of landward sand transport, overwash, and cut-through of new channels. The result is a widespread pattern of shoreline retreat or barrier migration toward the mainland. Lastly, the shape of the beach varies from year to year, season to season, and from hour to hour during storms, due to fluctuation in the other three factors.

The physical components of a typical barrier are as follows. The ocean side of a coastal barrier prior to development is normally characterized by a sandy beach or berm whose slope and width vary from one season to another according to weather, sea conditions, and sand supply. Sand bars just offshore form an integral part of the barrier sand exchange system. Similarly, sand dunes landward of the beach provide a "bank account" of sand to replenish the beaches at times of storm overwash. Behind the dunes on wider barriers lies a strip of fastland of variable width which may be wooded (shrublands or forests) or open grassland, depending upon local ecological conditions. The fastland gives way to coastal wetlands bordering the bay shore of the barrier. A coastal barrier is separated from the mainland by a bay, sound, or lagoon. If it is very close to the mainland, the intervening space may be entirely filled by tidal wetlands, as is common in New England (fig. 1.1).

AN OVERVIEW

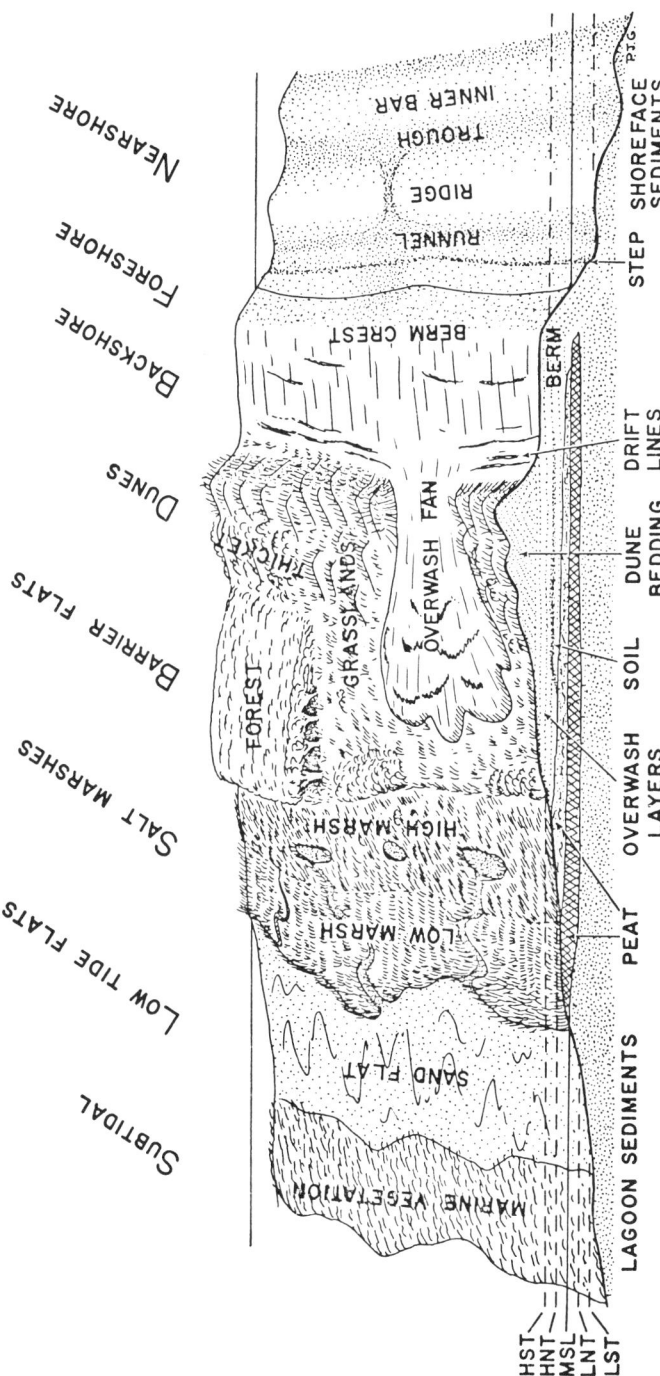

Figure 1.1. Cross-section of Typical Coastal Barrier from ocean at right to bay at left. Drawn by Paul J. Godfrey, University of Massachusetts, Department of Botany.

Coastal barriers protect both the mainland and the rich biotic habitats of the bay wetlands from the damaging effects of wave action during coastal storms and hurricanes. But barriers themselves are extremely hazardous during such events. Beaches and dunes are subject to drastic and sudden modification by storm waves, tides, and winds. In the event of a "storm surge"—an increase in normal tide levels by several feet—the barrier may be overwashed. This results in vast quantities of beach and dune material, and any manmade structures within reach, being transported toward and into the bay wetlands. Inlets or channels connecting the ocean and bay may be formed or reopened where storm waters erode the barrier, either during overwash or more likely when flood waters return to the sea.

Human Occupance of Coastal Barriers

Left to the play of natural forces, barriers are little affected by gradual migration. Beaches restore themselves from sand supply available in offshore bars, dune systems, and erodible uplands (e.g., glacial deposits). Vegetation recolonizes overwash areas and the bayside habitat flourishes, albeit in a slightly more confined spatial area. But the process of migration wreaks havoc with human attempts to occupy coastal barriers (Platt, 1985). Political and legal boundaries remain fixed in defiance of retreat of the shoreline. Buildings and lives are endangered by exposure to wave action, storm surge, high winds, and shore erosion. The natural instability of barriers is inimical to the social, economic, and political interests which seek to exploit them (see Furuseth and Ives paper).

Population in the United States has clearly been moving coastward in recent decades, although the exact rate and extent of coastal barrier demographic change is difficult to determine. At a very broad scale, the Bureau of the Census (1982, table 4) has calculated that in 1980, 118.4 million people (53% of the U.S. population) lived in 611 counties and independent cities within 50 miles of U.S. coastal shorelines, as compared with 60.5 million (46%) who lived in those jurisdictions in 1940. The American Meteorological Society (1975) has roughly estimated that six million people nationally live year-round within reach of a hurricane storm surge.

Precise estimates of demographic trends on coastal barriers are impeded by two factors. First, barriers are governed by a multitude of local authorities, some limited to the barrier and others including mainland areas as well. Unincorporated barrier communities are counted as part of the county in which they lie. It is impossible to determine from published census data exactly what proportion of the populations of counties and other mainland jurisdictions resides on coastal barriers. An inventory by Platt and Callahan (see paper in this volume) identified 234 political units governing portions of coastal barriers. A second complicating factor is the variation between year-round population and transient or seasonal

TABLE 1.1

LAND USE CHANGE ON ATLANTIC AND GULF COASTAL BARRIERS

	1945-55		1972-75		Amount of Change	
	acres	%	acres	%	acres	%
Urban/Built-up	90,410	5.5	228,679	13.6	+138,269	+153.0
Agricultural	14,746	0.9	10,160	0.6	-4,586	-31.0
Rangeland	101,019	6.1	98,812	5.9	-2,207	-2.0
Forest	168,161	10.2	152,224	9.1	-15,937	-10.0
Water Bodies	101,992	6.2	101,250	6.0	-742	-0.7
Wetland	918,015	55.6	838,882	50.0	-79,133	-9.0
Barrier Beach/ Dunes	256,357	15.5	249,241	14.8	-7,116	-8.0
TOTAL	1,650,700	100.0	1,679,248	100.0	+28,548	+2.0

SOURCE: Lins, 1980, table 27.

population; the latter is not counted in census data, but usually is several orders of magnitude larger than the permanent population.

Despite these drawbacks, West's paper in this volume computes population change between 1960 and 1980 in 156 municipalities which contain developed coastal barrier communities. Exactly what proportion of this change is attributable to the barrier beach portion of the community is unknown, as is the extent of seasonal variation from these year-round. Translation of demographic trends into estimates of land-use impacts is similarly fraught with pitfalls. Much new construction on coastal barriers in the form of hotels, motels, and seasonal condominiums is directed to that seasonal demand which is not counted in census data. Additional construction is non-residential in nature, such as restaurants, shopping centers, theaters, marinas, and convention centers.

Direct measurement of land-use change for all Atlantic and Gulf coastal barriers has been attempted in a U.S. Geological Survey study using aerial photography (Lins, 1980). This study revealed, as expected, a substantial increase in the proportion of developed land on coastal barriers during the period of (roughly) 1945 to 1975 (see table 1.1). Other categories of land use diminished accordingly, most notably wetlands which shrank by 9%, a loss of some 79,000 acres.

Modern remote sensing technology such as landsat and satellite data has not yet been applied to the refinement and updating of barrier land-use data at a national scale. There are, however, abundant examples of micro-studies of the impacts of human activities on barrier morphology, and, vice versa, as at Cape

Hatteras (Dolan, Godfrey, and Odum, 1973), Fire Island, New York (Psuty paper in this volume) and Ocean City, Maryland (Leatherman paper in this volume). Also, certain states such as Florida and Massachusetts have compiled detailed inventories of land use on their coastal barriers with funding assistance from the federal Coastal Zone Management Program.

Forms of Public Response

Structural Shore Protection and Beach Restoration

As population and investment at risk on coastal barriers have increased, so too have demands for public action to stabilize oceanfront shorelines to which most of the new development is oriented. Between 1936 and 1978, the U.S. Army Corps of Engineers assisted coastal communities in the construction of some 75 shore protection and beach restoration projects at an overall cost of $109 million (Ringold and Clark, 1980). Shore protection normally has taken the form of jetties or groin fields extending perpendicular to the shore to trap drifting sand, or sea walls in place of the vanishing beaches in front of developed properties. These measures, particularly seawalls, have hastened the loss of the beach due to increased scour and interference with access to replenishment sand. (See papers by Psuty, Nordstrom, Leatherman, and Gares in this volume).

The loss of beaches in front of major resort communities has led in turn to demands for expensive beach restoration projects. Between 1980 and 1982, the Corps restored a 10.5-mile, 300-foot-wide section at Miami Beach, Florida at a cost of $80 million (shared as follows: $10 million local, $20 million state, and $50 million federal). Proposals for comparable projects at Ocean City, Maryland and Sea Bright, New Jersey would cost approximately $33 million and $100 million respectively (Peters, 1985). Even where a beach is restored, it may be washed away again in subsequent hurricanes, as in the case of a Corps project at Rockaway Beach, New York. Even gradual erosion requires recurrent restoration projects.

Public Preservation

A different form of public response to coastal barrier development has been to preserve certain barrier beaches and associated ecosystems through public acquisition and management. The most extensive of these efforts has been the establishment of nine national seashores on the Atlantic and Gulf coasts (fig. 1.2). These facilities collectively include 454,000 acres of land including wetlands, of which 276,000 acres are located on coastal barriers (U.S. DOI, 1982, p. 14). The U.S. Fish and Wildlife Service administers 38 national wildlife refuges (twelve of them

AN OVERVIEW

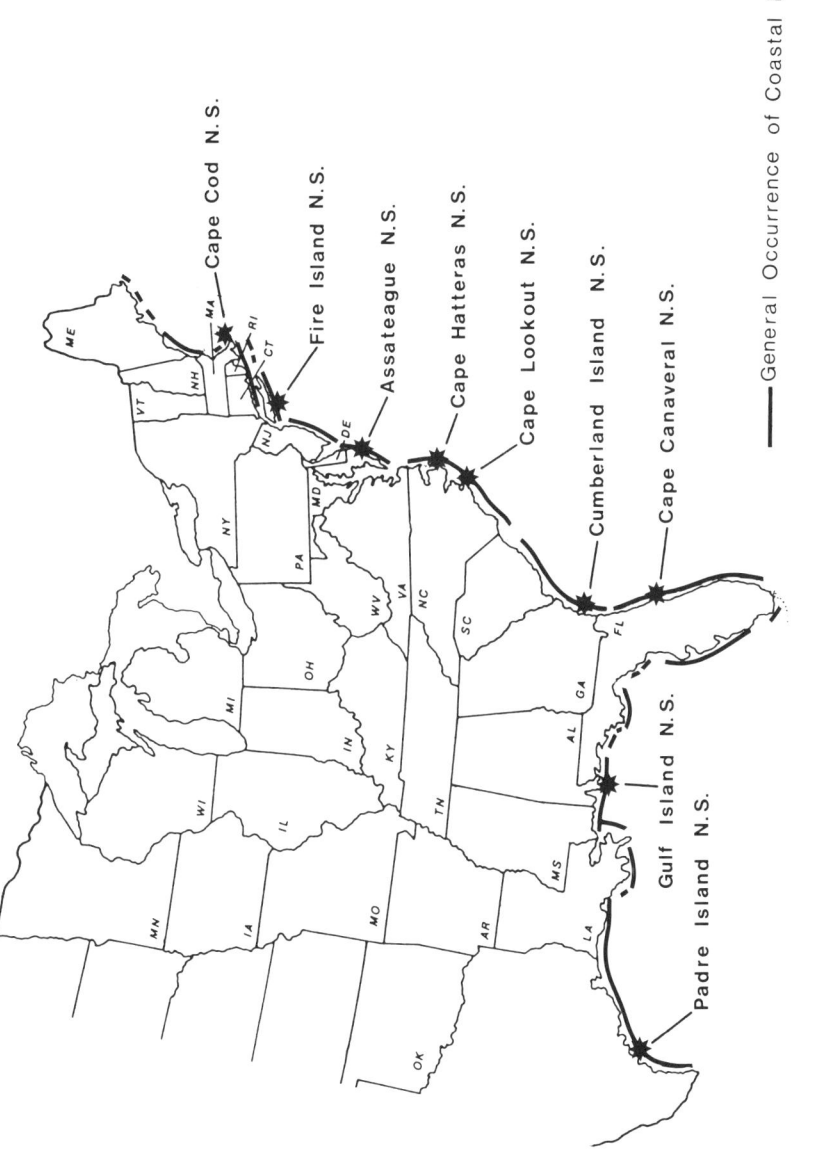

Figure 1.2

AN OVERVIEW

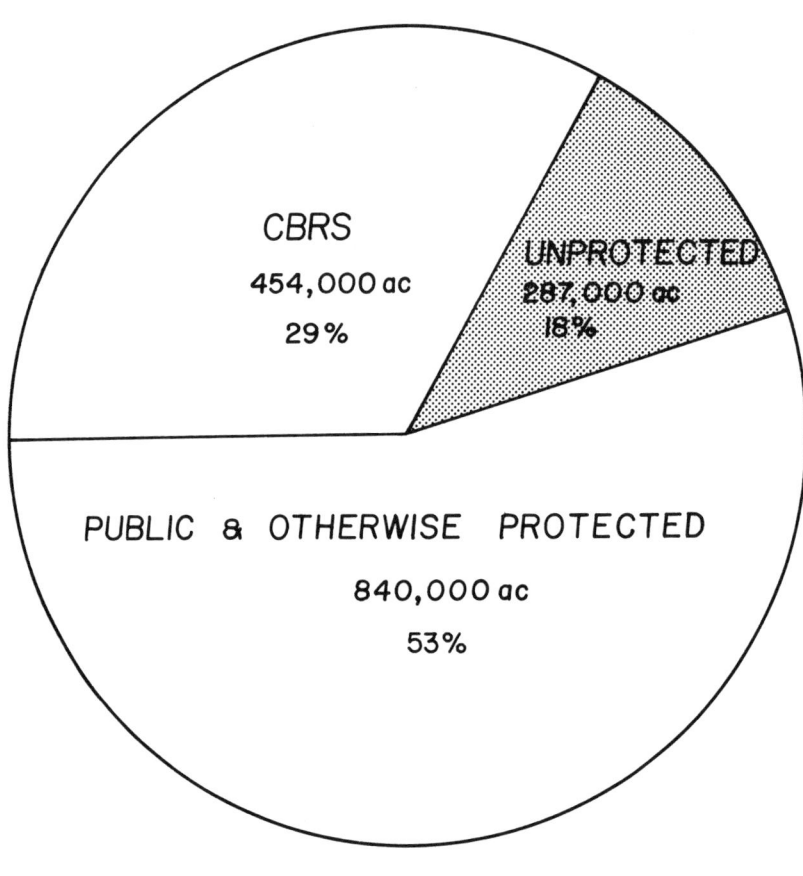

Figure 1.3

established since 1961) involving another 206,000 acres on coastal barriers, including another 50 miles of beachfront (Miller, 1981, p. 8).

Complementing these federal holdings are a number of smaller parcels owned and managed by states and local governments. By 1980, the federal Land and Water Conservation Fund had assisted at least 112 nonfederal acquisition projects on coastal barriers costing over $100 million. These projects were located on 74 different barriers in 17 states (U.S. DOI, 1980, p. vii). Still others have been protected through acquisition by The Nature Conservancy and other private land trusts. The Department of the Interior estimates that about 880 million acres or about 53% of all coastal barrier acreage is protected through public or quasi-public ownership and management (fig. 1.3) (Miller, 1981, p. 8).

Public preservation of coastal barriers encounters a pervasive problem in the form of incompatible activities on privately owned inholdings and adjacent lands which impinge upon the ecological, scenic, and recreational qualities of preserved areas. This problem, which is shared by public parks elsewhere, is particularly acute in the case of some of the national seashores such as Fire Island and Cape Hatteras, where pre-existing communities are intermingled with lands in public ownership. This results in a hodgepodge of jurisdictions and potential conflict in public policies and actions. Mantell and Duerksen address some of these issues in their paper.

Flood Insurance

The National Flood Insurance Program (NFIP), established by Congress in 1968, offers flood insurance coverage to owners of flood-prone land (both inland and coastal), if the community in which such land is situated adopts floodplain management regulations that satisfy NFIP standards. Flood insurance is a popular commodity in coastal communities. Although only 1,395 out of some 17,000 communities with identified flood risk areas are coastal, 70% of NFIP policies and 73% of coverage pertain to coastal communities. The success of the NFIP as a hazard mitigation tool depends upon several factors, e.g., the content of floodplain management regulations, the monitoring and enforcement of local regulations by federal and state authorities, the level of actuarial rates, and the accuracy of mapping of hazard areas (see Dawson paper).

Coastal Barrier Resources Act

Federal flood insurance has been identified as a stimulus to new construction and reconstruction in coastal hazard areas (U.S. General Accounting Office, 1982). Several other forms of federal assistance—e.g., highway and bridge subsidies, sewer and water grants, beach protection—have also been identified as further incentives to coastal development, particularly on barrier beaches (Miller, 1981).

Between 1979 and 1982, several bills were introduced in Congress to acquire or otherwise restrain growth on still-undeveloped coastal barrier acreage. A rider to the Omnibus Budget Reconciliation Act of 1981 (42 U.S.C. Sec. 4028) prohibited flood insurance for new or substantially improved structures on designated undeveloped barriers after October 1. This approach of denying federal subsidies for new development in designated areas was expanded by the Coastal Barrier Resources Act of 1982 (CBRA) (16 U.S.C. Secs. 3501-10), which received strong support from both environmentalists and the Reagan Administration. According to Kuehn (1984, p. 669):

> The unique intertwining of purposes such as minimizing threats to human life and harm to natural resources and curtailing extensive government financial involvement in development made CBRA extremely popular politically.

A key task was the identification of "undeveloped coastal barriers" within which federal development subsidies and flood insurance would be unavailable for new construction or substantial improvements. In 1982, a Barrier Island Task Force of the Department of the Interior had identified 188 such units totalling 750 miles of ocean-facing shoreline (U.S. DOI, 1982, p. 23). As adopted into law by the CBRA, the "Coastal Barrier Resources System" (CBRS) comprised 186 units totalling 656.18 miles of ocean shoreline or about 24% of the coastline of Atlantic and Gulf barriers. In area, the system amounted to 454,000 acres or about 29% of all barrier beach acreage (fig. 1.3). The CBRS is limited to privately owned portions of coastal barriers, characterized by development of no more than one manmade structure per five acres of fastland (see Godschalk paper).

Developed Coastal Barriers

The entire chain of Atlantic and Gulf coastal barriers amounts to 1.6 million acres of which half comprise wetlands (Lins, 1980). According to previous discussion, those areas which are publicly owned or "otherwise preserved" amount to about 840,000 acres. Barriers included in the CBRS account for an additional 454,000 acres. This leaves some 288,000 acres of other barrier areas which belong to neither of these two "protected" categories (fig. 1.3). These are "developed coastal barriers" whose surface area amounts to 18% of the total system, but —being ocean-oriented—whose shoreline amounts to 1,050 miles or about 40% of the total barrier shoreline. Developed coastal barriers include a wide diversity of communities and types of development, ranging from full-fledged cities like Atlantic City, Miami Beach, and Galveston to tiny unincorporated resort developments. As noted earlier, some 234 units of local and county government share jurisdiction over developed barrier communities (see papers by Mitchell and by Platt and Callahan). The CBRS threshold of one structure per five acres of fastland leaves much sparsely developed barrier acreage ripe for infilling with the benefit of federal subsidies.

This variegated collection of communities whose only common link is their location on coastal barriers was the subject of the "Cities on the Beach" conference and is thus the concern of this volume. Individual authors provide a variety of insights and recommendations on a broad array of barrier management issues. Those considered in this volume include recreational usage (Silberman; Janiskee and Lovingood), hurricane planning and evacuation procedures (Baker; Benedict; Jones and Spangler), financing of infrastructure (Miller), land-use planning and growth management (Brower and Beatley, Hamann, and Tripp).

The following express three broad conclusions which the speakers and authors seem generally to advocate:

1. Developed coastal barriers are extremely diverse in physical form, demography, political governance, and economic function. National and state policies regarding the management of such areas must reflect a recognition of this diversity and avoid pat generalizations and simplistic solutions which may be inappropriate in many cases.

2. Exposure of lives and investments to risk from natural hazards is increasing on most developed coastal barriers due to infilling of vacant land, intensification of already developed land, and rising property values of existing development. The federal government is continuing to contribute indirectly to this situation by providing the various benefits in developed areas that are now prohibited on undeveloped barriers. Furthermore, denial of such benefits in the CBRS actually reinforces pressures for further development on developed coastal barriers. The federal government must address the need to extend the CBRA approach to limit federal subsidies to at least the "high hazard" areas of developed barriers. In addition, local hazard mitigation measures such as dune preservation laws and improved evacuation capabilities must be encouraged.

3. Developed coastal barriers face a variety of urban planning problems aside from natural hazards which threaten the quality of life for residents and visitors alike. These include such problems as inadequate parking and beach access, inadequate or contaminated water supplies, visual blight through excessive use of signs and poor public design standards, urban decay, especially in older and larger coastal barrier communities, disposal of liquid and solid wastes, loss of remaining areas of natural habitat and open space, and inadequate housing, schools, and other services for residents or would-be residents of these communities (e.g., employees of restaurants and hotels). Each coastal barrier community is unique as to the exact nature and combination of such urban problems that it must address. National and state policies, however, should seek to promote habitability and environmental quality, as well as reduced exposure to hazard risks, on developed coastal barriers. These are, after all, the principal access points to the seashore for much of the American population. Cities on the beach are a fait accompli. We must find ways to optimize the pleasures of visiting or residing in them while minimizing the social and environmental costs of overbuilding them.

References

American Meteorological Society. 1975. "The Hurricane Problem." *Bulletin of the American Meteorological Society* 57: 8.

Barth, M.C., and J.G. Titus. 1984. *Greenhouse Effect and Sea Level Rise*. New York: Van Nostrand Reinhold.

Dolan, R., P. Godfrey, and W.E. Odum. 1973. "Man's Impact on the Barrier Islands of North Carolina." *American Scientist* 61 (2): 152-166.

Dolan, R., B. Hayden, and H. Lins. 1980. "Barrier Islands." *American Scientist* 68: 16-25.

Godfrey, P.J. 1976. "A Look at a Retreating Shoreline." *Trends* (July/Aug/Sept): 33-37.

Kaufman, W., and O.H. Pilkey. 1983. *The Beaches Are Moving: The Drowning of America's Shoreline*. Garden City: Anchor Press/Doubleday.

Kuehn, R.R. 1984. "The Coastal Barrier Resources Act and the Expenditures Limitation Approach to Natural Resources Conservation: Wave of the Future or Island Unto Itself?" *Ecology Law Quarterly* 11: 583-670.

Lins, H.F. 1980. *Patterns and Trends of Land-Use and Land Cover on Atlantic and Gulf Coast Barrier Islands*. USGS Prof. Paper no. 1156. Washington: U.S. Government Printing Office.

Miller, H.C. 1981. "The Barrier Islands: A Gamble with Time and Nature." *Environment* (November): 6-12; 36-41.

Peters, J.W. 1985. "War of the Waves." *The Baltimore Evening Sun*. (Special reprint).

Platt, R.H. 1985. "Congress and the Coast." *Environment* 27 (July/Aug): 12-17; 34-40.

Ringold, P.L., and J. Clark. 1980. *The Coastal Almanac*. San Francisco: W.H. Freeman and Company.

Schwartz, M.L. (ed.). 1982. *The Encyclopedia of Beaches and Coastal Environments*. New York: Van Nostrand Reinhold.

U.S. Bureau of the Census. 1982. *Statistical Abstract of the United States—1982-1983*. Washington: U.S. Government Printing Office.

U.S. Department of the Interior. 1980. *Alternative Policies for Protecting Barrier Islands along the Atlantic and Gulf Coasts of the United States and Draft Environmental Impact Statement*. Washington: DOI.

U.S. Department of the Interior. 1982. *Undeveloped Coastal Barriers: Report to Congress*. Washington: U.S. Government Printing Office.

U.S. General Accounting Office. 1982. *National Flood Insurance: Marginal Impact on Floodplain Development*. (CED-82-105). Washington: GAO.

THE COASTAL BARRIER RESOURCES ACT

THE 1982 COASTAL BARRIER RESOURCES ACT: A NEW FEDERAL POLICY TACK

David R. Godschalk

University of North Carolina at Chapel Hill

Introduction

The Coastal Barrier Resources Act of 1982 (CBRA) reflects the recent conservative trend as it applies to environmental policy. A unique marriage between fiscal conservativism and environmental conservation, CBRA sets a new course to the right by withdrawing, rather than adding, federal incentives in order to achieve public purposes.

CBRA expresses a federal policy not to subsidize future development on hazardous, undeveloped coastal barriers. It does not prohibit owners from building there, but it does shift the cost of infrastructure and the risk of loss from the federal treasury to the private sector or to state or local governments. Its purpose is to minimize loss of life, wasteful expenditures of federal revenues, and damage to natural resources.

An initial study indicates that the aims of shifting infrastructure costs and risks of loss away from the federal treasury are being achieved, at least in some areas (Godschalk, 1984). Whether the aims of conserving natural resources and minimizing loss of life will also be achieved is less clear. This will depend upon the extent to which state and local governments reinforce CBRA goals, private insurance companies make flood insurance available, banks and development finance agencies relax insurance conditions on future loans and mortgages, private development companies modify plans for coastal barrier projects, and conservation organizations redraw priorities for acquiring coastal barrier open space.

Bold and elegant, the CBRA strategy is highly selective, attempting to achieve broad public purposes by indirectly influencing the market for development on undeveloped portions of coastal barriers. By limiting federal actions only to withdrawing future federal flood insurance and financial assistance and by limiting the focus only to designated undeveloped areas (less than a quarter of the total ocean-facing barrier shoreline), CBRA focuses on those actions most accessible to federal expenditure manipulation and those areas least likely to have vocal political constituencies. Many important issues remain unaddressed by CBRA, however, highlighting the need for a comprehensive barrier management policy that incorporates other critical development influences and the remaining three-quarters of the coastal barrier shoreline.

The Act's Provisions

Congress declared in the CBRA that past federal government actions that encouraged development on coastal barriers resulted in loss of resources, threats to life and property, and expenditures of millions of tax dollars every year. To reverse this situation, it designated a chain of 186 "undeveloped" areas stretching from Maine to Texas as the "Coastal Barrier Resources System." In such areas, CBRA prohibits further federal flood insurance coverage as well as federal funding for bridges and roads, utilities, access channels, erosion control, storm protection, community development, post-storm redevelopment, and non-emergency disaster relief. Flood insurance was terminated on October 1, 1983; the other curtailments took effect on October 18, 1982, the date of passage of the Act.

Coastal barriers are defined by the Act as depositional geologic features consisting of unconsolidated sedimentary materials shaped by wave, tidal, and wind energies. Landward aquatic habitats protected by barrier beaches and dunes from direct waves are included in the definition of resources subject to the Act. Undeveloped barriers are those with few artificial structures and no significant human impediments to geomorphic and ecological processes.

Administration of CBRA is assigned to the Department of the Interior (DOI), which maintains and periodically reviews the maps of designated areas within the system. Initial inventory and mapping of undeveloped barriers was performed by DOI pursuant to the Omnibus Budget Reconciliation Act of 1981 (U.S. Department of the Interior, 1982). Criteria for units identified in this study included: (1) less than one walled and roofed building per five acres of fastland, (2) absence of urban infrastructure (vehicle access, water supply, wastewater disposal, and electrical service to each lot), and (3) not part of a development of 100 or more lots. In general, the minimum portion considered for designation contained a quarter mile of ocean-facing shoreline and extended from the beach to the bay side.

Congress modified DOI's initial maps after discussions with affected landowners and interest groups. A Senate committee report stated that the criteria for definition in the text of the Act are advisory only; the maps adopted as part of the Act would be the last word as to which areas are included in the Coastal Barrier Resources System. In a process that blended politics with correction of technical errors, Congress reduced the undeveloped areas from 721 to 656 miles of ocean-facing shoreline prior to adoption. To further change the areas covered will require an amendment to the law. In a report to Congress in 1985, DOI

TABLE 2.1

CLASSIFICATION OF COASTAL BARRIER SHORELINE, 1982

Type	Shoreline	
	mi.	%
Undeveloped	656	24
Protected	979	36
Developed	1050	40
Total	2685	100

SOURCE: U.S. DOI, 1982.

recommended major expansion of the Coastal Barrier Resources System to include "otherwise protected" areas (those managed by public or private conservation organizations), related coastal landforms such as keys and mainland beaches, and Great Lakes, Pacific Coast, Alaskan, Hawaiian, Puerto Rican, Trust Territories and Virgin Island barriers.

A summary of the amount of ocean-facing shoreline by types illustrates the distribution of areas along the Atlantic and Gulf coasts as originally defined by CBRA in 1982 (table 2.1).

Coastal Barriers Institutional System

Impacts on conservation and development stemming from CBRA will depend upon the responses of a variety of public and private groups. These groups, their responses, and the resulting impacts can be visualized as a "coastal barriers institutional system." As diagrammed in figure 2.1, this system is set in motion by the cutoff of federal flood insurance and financial assistance to the undeveloped coastal barriers.

Five groups are faced with choices critical to conservation and development outcomes within their operating areas:

1. *State and local governments* must decide whether or not to reinforce CBRA objectives in their coastal management programs. Especially significant will be decisions concerning public spending on infrastructure and open space, and on land-use planning and development regulation.

2. *Private insurance companies* face the choice of whether or not to make privately underwritten flood insurance available to replace withdrawn federal flood coverage. Important questions include whether private flood insurance will

20 THE COASTAL BARRIER RESOURCES ACT

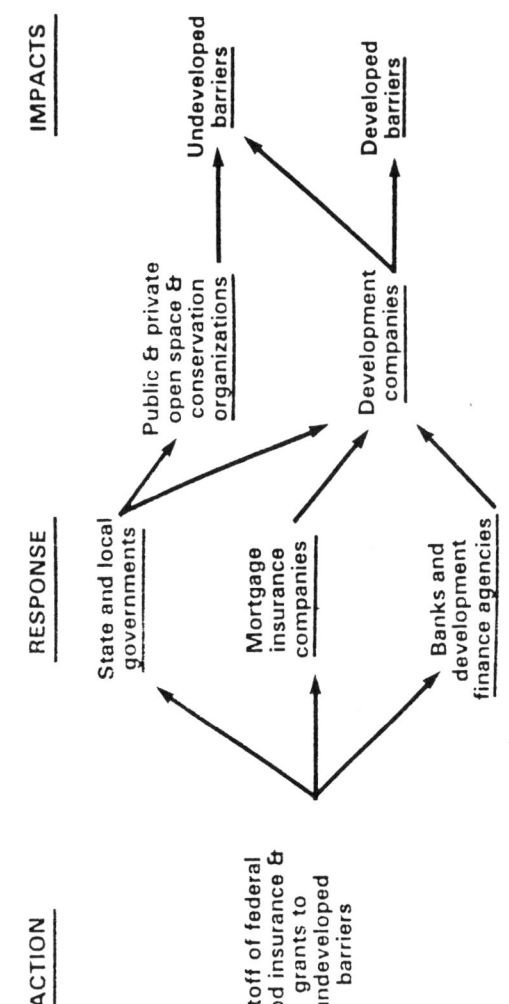

Figure 2.1. Coastal Barriers Institutional System

be available from admitted (state regulated) or non-admitted companies (such as Lloyds of London), at reasonable or very large premiums and/or deductibles, or only to condominium or apartment projects. If private flood insurance does become generally available, will it be withdrawn following major coastal storms or hurricanes?

3. *Banks and development finance agencies* must decide how to replace required federal flood insurance as a precondition for offering a loan or mortgage. Since CBRA withdraws both federal flood insurance and federal loans and mortgages through the Federal Housing Administration, Veteran's Administration, Small Business Administration, and Federal Home Loan Administration, the role of private capital providers becomes crucial.

4. *Public and private open space and conservation organizations* must decide whether to acquire more land on undeveloped coastal barriers. They must ask if CBRA restrictions, in concert with state and local government actions, are sufficient to conserve the undeveloped coastal barriers or are major conservation investments also needed? Or should they focus their land acquisition efforts on developed barrier areas, under the assumption that these are more threatened by development pressure?

5. *Private development companies* must decide how to respond to CBRA restrictions and related actions. Developers without major land holdings on undeveloped coastal barriers are likely to be cautious about investing there. Developers with major land holdings in the designated areas can either seek to sell their land or develop it without use of federal flood insurance and infrastructure assistance. Given the cost of oceanfront land, such developers are likely to focus on high-density, luxury condominium projects, where they can make a profit on their investments even with higher infrastructure costs and where homeowners' associations can deal with negotiating insurance contracts and maintaining infrastructure.

Probable Impacts of CBRA

The Department of the Interior anticipates that CBRA will save the federal government approximately $5.4 billion over the next 20 years and will also enhance protection of coastal barrier floodplains and wetlands (U.S. Department of the Interior, 1983, pp. II-24-26). On the other hand, CBRA could generate pressure in existing coastal communities for increased density, infilling of vacant land, expansion of infrastructure, and even demolition and redevelopment of older, single-family residences to make way for new, multi-family projects. While these spillovers can bolster the economies of developed coastal areas, they can also strain the capacities of small communities to plan and manage new growth.

Since withdrawal of flood insurance took effect in October of 1983, only limited experience with the effects of CBRA has been documented. One study has assessed initial impacts in three states where major impacts are expected—North and South Carolina and Florida—from the viewpoints of developers, government officials, and conservationists (Godschalk, 1984). Another study has looked at the response of the private insurance industry in South Carolina (Griepentrog, Gaines, and Schroath, 1984). DOI has conducted four regional studies and several analyses of other aspects of CBRA, such as taxation and regulation (U.S. Department of the Interior, 1985). In addition, a U.S. District Court has upheld CBRA's designation of land on Topsail Island, North Carolina, as undeveloped (*Bostic v. United States,* 581 F. Supp. 254 (E.D.N.C. 1984)).

These sources concur that the initial impacts of CBRA, while not pronounced, generally fulfill the Act's intentions, although unexpected results could occur in the future. Affordable private flood insurance has not yet become widely available to replace withdrawn federal flood insurance. Griepentrog, Gaines, and Schroath (1984) predicted that the private insurance market would fill the void left by CBRA actions, but a downturn in the underwriting cycle has postponed a widespread replacement. Meanwhile, hurricanes Alicia in Texas (1983), Diana in North Carolina (1984), and Gloria in the Northeast (1985) have been uncomfortable reminders of potential coastal hazards to private insurers.

Replacement of infrastructure funding by state and local governments also has not occurred on a widespread basis, due in part to the same fiscal stringency and conservatism that fostered CBRA. The majority of the 68 respondents to the Godschalk survey (1984) viewed developers as the primary source of replacement infrastructure financing. Meanwhile, the governors of Florida (Executive Order 81-105, 1981) and Massachusetts (Executive Order 181, 1980) have issued policies that prohibit state funding of barrier infrastructure.

Case Studies

Topsail Island

Issues come into sharper focus when specific barrier situations are examined, as in the case studies of Topsail Island, North Carolina and Hutchinson Island, Florida conducted by the author (Godschalk, 1984). Topsail Island is a barrier island off the southern coast of North Carolina near the Camp Lejeune Marine Base, where the withdrawal of federal flood insurance apparently has had a significant initial impact on development (fig. 2.2). Already supplied with a complete infrastructure base, the northeastern part of the island is designated undeveloped under CBRA. Developers of a 1,200-acre area, including over a mile of ocean frontage, were joined by other developers in a civil suit alleging that their land was erroneously designated as undeveloped and requesting an injunction against terminating federal assistance. In January 1984, the U.S. District Court ruled against them, holding that the designation was rationally justified (*Bostic v. United States,* 581 F. Supp. 254 [E.D.N.C. 1984]). The decision in this case, the

THE COASTAL BARRIER RESOURCES ACT   23

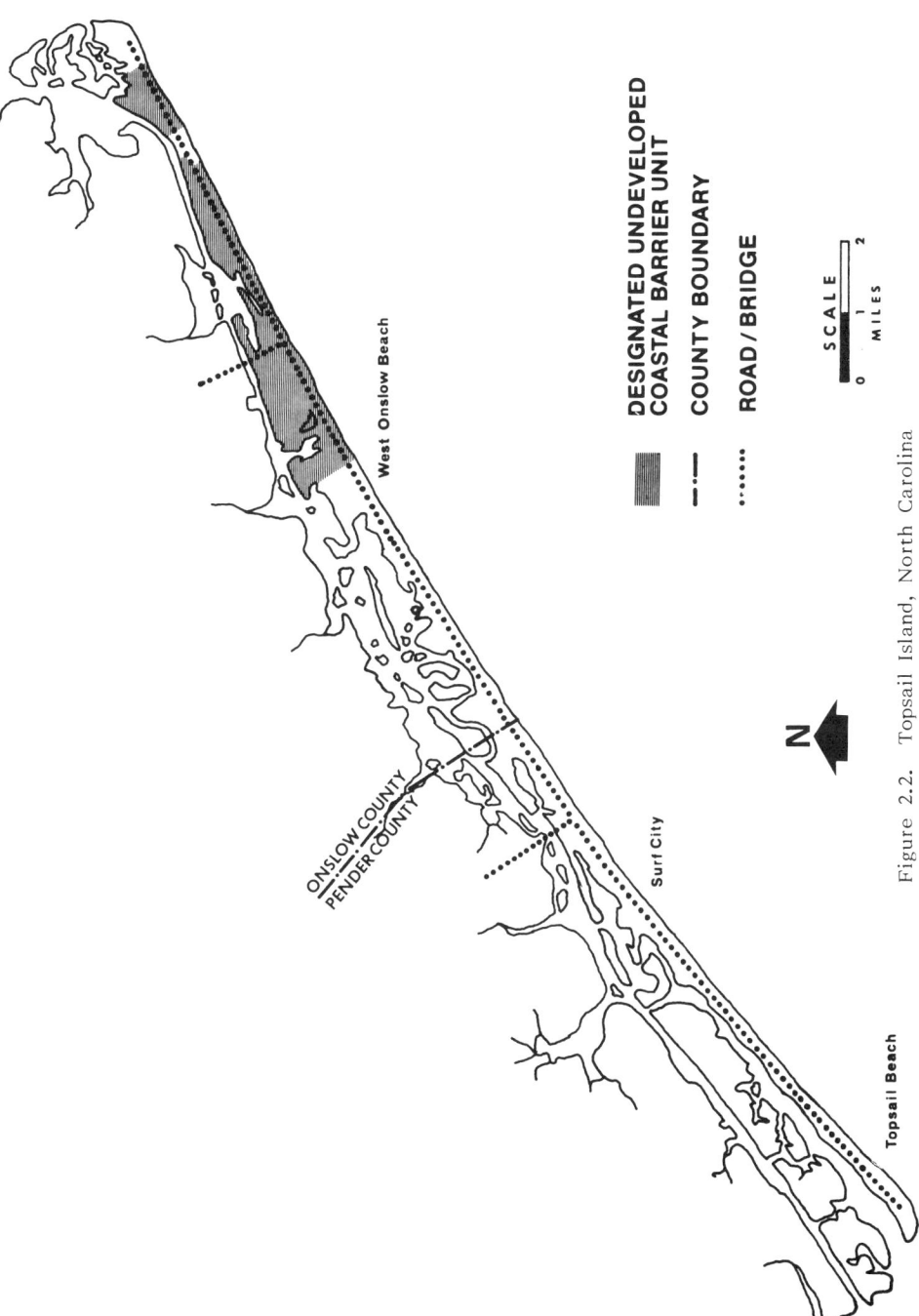

Figure 2.2. Topsail Island, North Carolina

first legal challenge to CBRA, was affirmed by the U.S. Court of Appeals in February of 1984 (*Bostic v. United States,* 753 F.2d 1292 [4th Cir. 1985]).

No private flood insurance replacement had been obtained for any of the Topsail Island projects when the lawsuit was filed in early 1984. Both developers and insurance brokers, however, believed that replacement insurance eventually would be obtained for multi-family, condominium projects but were unsure about replacement insurance for single-family or duplex units. (Private insurance, including flood damage coverage, had been obtained for the North Topsail projects by 1985.) Previously, for large projects, federal flood insurance provided the first layer of coverage, up to its limit, and private insurance with a deductible equal to the federal limit provided the second layer. Single-family and duplex dwelling units also might have such two-layer coverage, depending on whether their value significantly exceeded the federal limit.

Hutchinson Island

Hutchinson Island is a Florida barrier island where the withdrawal of federal assistance for infrastructure, in combination with state and local government actions, had a significant initial impact on development (fig. 2.3). Lying off the Atlantic Coast adjacent to Fort Pierce, the middle section of the island is designated as undeveloped. An interesting anomaly is the location of a Florida Power and Light Company nuclear power plant between two of the undeveloped areas.

While Hutchinson Island has an infrastructure base, it is inadequate to serve the proposed high level of development. Bridge capacity is the most critical need. St. Lucie County imposed a moratorium on development project approvals through early 1984, after discovering that neither federal nor state funding would be available for new bridge construction. Florida Governor's Executive Order 81-105 declares that state funds are not to be used to subsidize growth or post-disaster redevelopment on hazardous coastal barriers. Developers sued the county and won a reversal of the moratorium, but the county appealed the decision and the suit was settled out of court. (*St. Lucie County v. North Palm Development et al.*, Fourth District Court of Appeals of Florida, No. 83-1863 and 83-2233). Meanwhile the state entered negotiations over the growth management conditions under which it might be willing to participate in financing a new bridge to meet existing needs.

Both case studies indicate that CBRA is being taken seriously. Even though most argue that the market eventually will find ways to accommodate the continuing demand for barrier island development, even without federal flood insurance and financial assistance, initial impacts seem to have slowed development. The all-important, long-term impacts, however, remain to be seen.

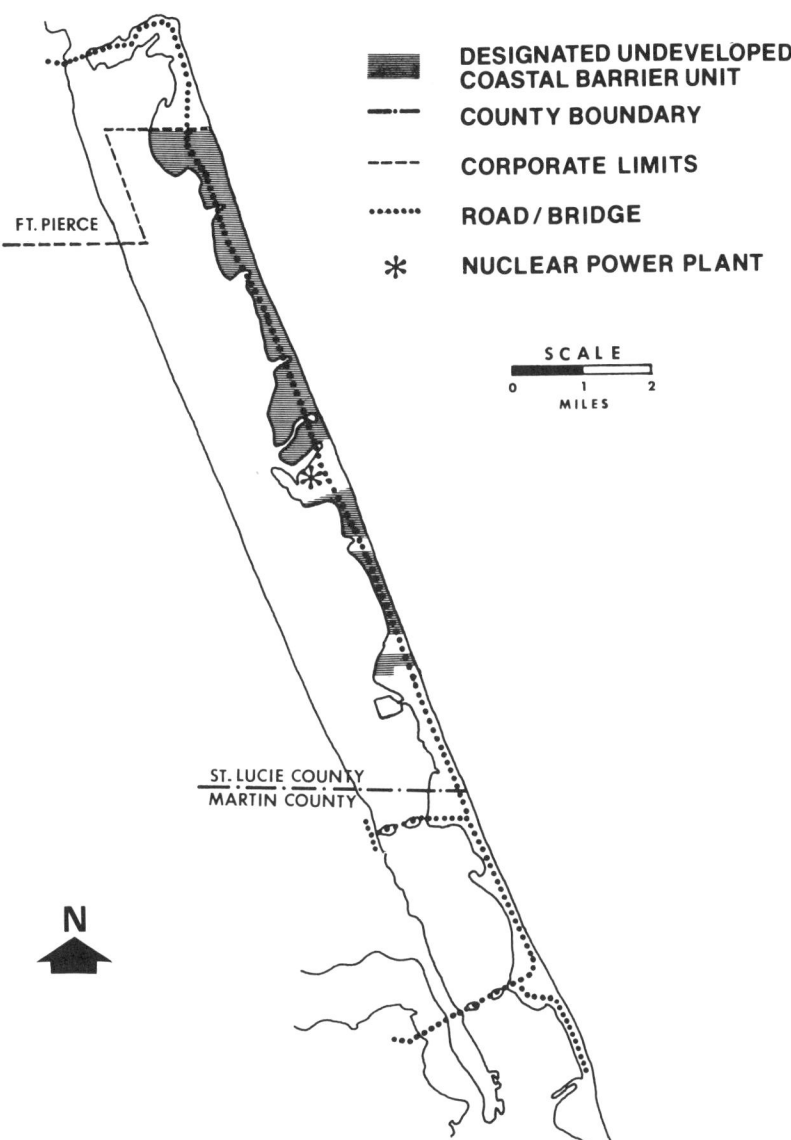

Figure 2.3. Hutchinson Island, Florida

Charting Long-Term Coastal Policy

CBRA is a move in the right direction. Removing federal development incentives in coastal barrier areas will reduce federal costs while simultaneously reducing public subsidies to barrier island development.

CBRA's limited actions, however, are not enough to ensure long-term conservation of the coastal barriers, even if geographic coverage is expanded in accordance with DOI's recent recommendations. Among likely future impacts are:

1. Additional high-density, expensive development on the undeveloped barriers, as the market seeks to overcome federal insurance and infrastructure withdrawal by moving to large condominium projects

2. Additional infill and redevelopment growth pressure on developed barriers, due to land supply constraints imposed by CBRA

3. Reduction in both single-family and moderate-cost housing on coastal barriers, due to increased private insurance and construction costs

4. Further legal and political attempts to overturn CBRA constraints, based on responses to future hurricane disaster relief needs, constitutional challenges, and economic development arguments

At this time, these further impacts are speculative. However, it is clear that CBRA is only a partial answer to barrier conservation and development management, leaving a number of unresolved issues to be dealt with by a broader public policy framework.

A more comprehensive coastal barrier policy would spell out the principles under which state and local management efforts, as well as private conservation and development initiatives, could be coordinated with federal programs. It would recognize and provide policies covering the full spectrum of existing barrier areas, including partially developed, as well as undeveloped areas (as in the proposed Florida rule implementing Executive Order 81-105). It would bring to bear a full slate of implementation tools, including taxes and regulations. It would regularly monitor and evaluate the impacts of public policy on barrier island conservation and development.

The new tack in coastal policy has worked well as a first step. Subsequent Congressional review should look further along the course and make the adjustments necessary to achieve an effective and comprehensive coastal policy.

References

Godschalk, D.R. 1984. *Impacts of the Coatal Barrier Resources Act: A Pilot Study.* Washington: Office of Ocean and Coastal Resource Management, U.S. Department of Commerce.

Grienpentrog, G.L., J.P. Gaines, and F.W. Schroath. 1984. *A Projection of the Commercial and Private Development to Result from the Changes in Availability and Cost of the Federal Flood Insurance Program.* Charleston: South Carolina Sea Grant Consortium.

U.S. Department of the Interior. 1982. *Undeveloped Coastal Barriers: Report to Congress.* Washington: Department of the Interior.

U.S. Department of the Interior. 1983. *Final Environmental Statement: Undeveloped Coastal Barriers.* Washington: Department of the Interior.

U.S. Department of the Interior. 1985. *Draft Report to Congress on the Coastal Barrier Resources System.* Washington: Department of the Interior.

THE GEOGRAPHICAL CHARACTERISTICS OF COASTAL BARRIERS

A MANAGEMENT-ORIENTED, REGIONAL CLASSIFICATION OF DEVELOPED COASTAL BARRIERS

James K. Mitchell

Rutgers - The State University
of New Jersey

Introduction

Despite a paucity of data, the one or two measures of "development" on coastal barriers that are generally available go a long way toward defining a regional framework that is useful for conceptualizing developed barrier management strategies. Only barrier beaches on the Atlantic and Gulf coasts that satisfy criteria used in the Coastal Barrier Resources Act (1982) (CBRA) are considered here. It is recognized that additional barriers occur elsewhere in the United States and that there exist geological features which are not technically coastal barriers but perform related functions and suffer comparable problems, such as the Florida Keys.

Principles of Coastal Barrier Management

The chairman of the Coastal Barriers Task Force correctly observed that simplicity and consistency are the principal strengths of the Coastal Barrier Resources Act (CBRA) (Davidge, 1983, p. 313-316). As a result of CBRA and related open-space initiatives, most of the *undeveloped* coastal barriers are now subject to a common set of clear-cut public policies designed to limit future development (U.S. Department of the Interior, 1983).

Simplicity and consistency should also be important guiding principles of any management program that is eventually adopted for *developed* coastal barriers. Unfortunately, the task of fashioning a developed barrier management program will be difficult for at least two reasons. Paradoxically, despite a wealth of case studies and anecdotal evidence, much less is known about human-modified environmental processes on developed coastal barriers than about less disturbed, undeveloped barriers. Moreover, what is known suggests that the issues and problems of developed barriers are so varied and complex that a CBRA-like management approach—centralized, uniform, and specific—is probably not feasible.

Coastal Barrier Data Bases

Data inadequacies hamper any detailed analysis of the human dimensions of developed barriers. Much of the necessary demographic, socioeconomic, land-use, and hazards data are either missing, incomplete, or inconsistent. For example, depending on the source consulted and definition applied, the number of identified coastal barriers in Massachusetts varies from 21 (U.S. Department of the Interior, 1980), to 27 (Lins, 1980), to 77 (U.S. Department of the Interior, 1983), to 675 (Smith, 1982). There are also many discrepancies with respect to the definition and extent of public ownership. The final environmental impact statement in support of CBRA (U.S. Department of the Interior, 1983, p. IV-14) notes that 29% of the adjacent beach frontage on New Jersey's coastal barriers is protected against development by virtue of being publicly owned. The state's shore protection master plan, however, lists more than three quarters (76.2%) of the beach frontage as publicly owned (New Jersey Department of Environmental Protection, 1980, pp. 1-39).

Three examples of systematic data inadequacies are particularly significant. First, very little is known about the size and composition of seasonal populations, especially in the northeastern states. Analysts believe that peak summer populations can range up to 20 times the permanent year-round totals recorded by the U.S. Census. New Jersey's developed coastal barrier population is approximately 186,000, but estimates of peak summer population range from 700,000 to 1.6 million (Mitchell, 1984, p. 23). Lack of information about seasonal populations hampers evaluation of safety hazards and management priorities. Second, the most complete land-use data sets describe only *areas* devoted to generalized uses, not *intensities* of use or specific uses (Lins, 1980). They do not distinguish among high-rise apartment and hotel complexes, scattered summer cottages, lagoon homes on dredged-access canals, or many other types of development with different management implications. Third, comprehensive data about natural hazard losses on barriers either have not been compiled or cannot readily be extracted from existing records because classification systems are based on communities rather than the barrier landforms within them (Federal Emergency Management Agency, 1984).

A detailed analysis of developed coastal barriers must patch together data from a variety of sources, including CBRA working documents, national overviews of specific coastal problems, state coastal management program studies, data on the National Flood Insurance Program, regional hurricane evacuation plans, and sources that focus on groups of coastal barriers. Prospective managers will benefit from more accurate, specific, consistent, and comparable information about developed coastal barriers. For now, existing, piecemeal data sources must suffice.

What is known about developed coastal barriers clearly indicates that they differ from undeveloped barriers in several important ways. Their disturbed ecosystems are less well understood by scientists and less sought after by conservationists. They are also beset by problems largely absent from undeveloped barriers. These include: pervasive fresh water deficits, salt water intrusion, and water pollution; traffic congestion; seasonally-based, local economies; underfinanced municipal governments and inadequate public services. These circumstances are complicated by divergent goals between year-round occupants and temporary residents and by lack of coordination among governments that share common coastal barriers (see Platt and Callahan, this volume).

At first glance, developed barrier management and coastal zone management appear synonymous. Both involve many competing demands for the use of a limited, sensitive, and hazardous environment (Mitchell, 1982). However, although developed barriers are small, they are not a microcosm of the entire coastal zone. They are a distinctive subset of the coastal zone, with sufficient common characteristics and problems that strong national or multi-state management initiatives are suggested.

Common Characteristics of Developed Coastal Barriers

The common characteristics of developed barriers are several. They are mostly given over to amenity housing, beach-oriented recreation, and related services. Most lack significant industrial plants, commercial ports or similar facilities, although small offshore support bases and like uses are occasionally found along the Gulf Coast. They contain little agricultural or forest land. By definition—and unlike the wetlands they often protect—they host few pristine, natural ecosystems.

Most important, what distinguishes developed barriers from other coastal areas and links them to undeveloped barriers is the fact that they contain a distinctive natural sedimentation system. The hallmark of this system is rapid landform change in response to natural processes that also threaten human life and property. Storms, erosion, floods, and other hazards are of overriding importance to the evolution, maintenance, and destruction of developed barriers and to the people who occupy them. As such, a developed barrier management program must be—first and foremost—a program that addresses natural hazards.

Natural Hazard Losses

It is widely believed that developed coastal barriers suffer disproportionately large hazard losses compared to most parts of the United States. Although probably true, that assertion is not easy to substantiate with quantitative evidence. Certainly hurricanes and other storms have inflicted significant damage on Galveston, Gulf Shores, Westhampton Beach, Scituate, and other developed barriers, but it is difficult to disentangle the specific magnitude of impacts on coastal barriers from those that occurred in adjacent areas. Analysis of National Flood Insurance Program reimbursements provided over the past seven years illustrates the somewhat ambiguous conclusions that can emerge from studies of developed barrier losses (table 3.1). Relative to their populations, barrier island communities are paid disproportionately larger claims than mainland communities. The level of reimbursement ($5,869) is very close to the average for most hazard-prone areas ($5,935), but substantially less than the worst-affected communities. Houston, Mobile, and Baytown, for example, have received a comparable number of substantially larger flood insurance payments (averaging $10,249). Perhaps differences in the value of commercial property at risk explain these contrasts.

Hazard losses on coastal barriers are large and growing. What accounts for this trend? The probability of occurrence of storms and floods is not changing, so increased losses may be due to the fact that coastal barriers become more vulnerable as they become more developed. Therefore, it is appropriate to analyze the degree to which developed barriers are in fact "developed."

Developed Coastal Barriers: A Profile

It is estimated that 1.7 million acres of coastal barriers are distributed along 2,685 miles of the U.S. Atlantic and Gulf coasts (U.S. Department of the Interior, 1983). In the mid-1970s, when the most recent comprehensive land-cover data were compiled, 14% of that area and 37% of the ocean frontage were occupied by buildings, roads, and related features (Lins, 1980) (table 3.2). These are the developed coastal barriers.

Preferred Oceanfront Sites

The disproportionate development of oceanfront locations stems largely from public preferences for homesites with direct access to beaches and water. This has two significant management implications. First, strip development with subsequent infilling has historically dominated the land conversion process on coastal barriers to the point where available acreage becomes predominantly built over. Compared with areally compact patterns of settlement, the provision of public-hazard protection works, other infrastructure, and public services is inefficient and costly. The exposure of adjacent undeveloped land to potential environmental disruption is correspondingly greater. Second, narrow coastal barriers tend to become more heavily developed than wider barriers. For example, the predominantly narrow

TABLE 3.1

FLOOD INSURANCE REIMBURSEMENTS FOR SELECTED BARRIER ISLANDS
(1978-1984)

State	Community	Population (1980)	Number of Reimbursements	Value of Reimbursements
New York	Westhampton Beach	NA	278	$ 1,431,014
New Jersey	Sea Bright	1,812	305	1,364,042
	Monmouth Beach	3,318	236	1,066,752
	Long Beach Twp.	3,488	247	462,831
	Brigantine	8,318	378	1,018,932
	Atlantic City	40,199	611	2,125,026
	Ocean City	13,949	722	1,953,722
	Sea Isle City	2,644	406	1,258;350
	Avalon	2,162	274	616,478
	North Wildwood	4,714	591	1,763,279
	Wildwood	4,913	289	802,581
	West Wildwood	360	385	1,688,394
Florida	Longboat Key	4,843	252	752,560
	St. Petersburg Beach	9,354	221	452,013
	Treasure Island	6,316	280	1,861,157
	Madeira Beach	4,520	323	882,338
	Pensacola Beach	NA	752	4,134,689
Alabama	Gulf Shores	NA	792	16,226,197
Louisiana	Grand Isle	NA	370	968,420
Texas	Galveston	61,902	3,235	16,206,561
	South Padre Island	NA	820	7,379,691
			10,975	$ 64,415,027
			Average payment:	$5,869
250 Communities with Largest number of reimbursements:				
			221,727	$1,316,105,799
			Average payment:	$5,935
Sample Non-Barrier Communities				
Texas	Houston	1,595,138	7,967	$ 71,158,053
	Baytown	56,923	1,396	21,987,507
Alabama	Mobile	200,452	2,266	26,041,284
			11,629	$ 119,186,844
			Average payment:	$ 10,249

SOURCE: FEMA, 1984.

TABLE 3.2

DEVELOPED LAND ON COASTAL BARRIERS BY STATE, 1980

STATE	AREA		FRONTAGE	
	Acres Built Up	% Barrier Area Built Up	Miles of Ocean Front Developed	% Ocean Front Developed
Maine	1,165	72%	15.8	56%
New Hampshire	780	72	8.1	100
Massachusetts	8,128	22	47.7	22
Rhode Island	1,226	35	11.2	30
Connecticut	576	42	8.8	41
New York	11,578	35	76.4	44
New Jersey	22,719	47	73.3	69
Delaware	2,956	29	9.7	20
Maryland	1,848	14	8.8	29
Virginia	1,144	2	19.7	18
North Carolina	21,625	14	119.3	37
South Carolina	13,081	8	63.5	42
Georgia	8,436	5	16.0	15
Florida	101,988	20	460.7	63
Alabama	5,273	19	30.1	51
Mississippi	0	0	0	0
Louisiana	6,746	18	14.9	10
Texas	19,410	5	64.2	18
TOTAL	228,679	14%	1,048.2	37%

SOURCE: H.F. Lins, 1980.

barrier beaches from Delaware to Maine are more than three times as heavily built up (37.1% occupied) as the generally wider barrier beaches of the southern states (11.6%). Among northern barriers, narrower islands are also more heavily settled than their wider neighbors. Inasmuch as narrow barriers tend to possess lower dunes and are more vulnerable to storm surge, overwash, and breaching, they are particularly hazardous locations for buildings. Conversely, unless they are very densely settled or far removed from safe mainland areas, narrow barriers may be somewhat easier to evacuate, because their populations are significantly limited by the scarcity of buildable land.

Alternative Development Processes

Clearly, the trend toward increasing vulnerability is not homogeneous. Different development processes create different types of vulnerability. For example, many developed coastal barriers are seasonally occupied, but an increasing number host permanent populations composed of retirees or workers who commute to jobs in nearby metropolitan centers (e.g., Long Beach, New York). Both types of community are vulnerable to natural hazards, but the mix of threatening events (e.g., hurricanes *vs.* northeasters), properties at risk (e.g., apartments *vs.* single-family homes), and preferred responses (e.g., evacuation *vs.* sheltering) is likely to be quite different for each. The contrasts between places like Scituate, Massachusetts and Wrightsville Beach, North Carolina are illustrative. Some coastal barriers are being developed for the first time (e.g., Marco Island, Florida), whereas others have undergone successive cycles of redevelopment, either as a result of repeated natural disasters (e.g., Westhampton Beach, New York) or the replacement of decaying structures via urban renewal projects (e.g., Atlantic City, New Jersey). Often land is first converted by individuals or small firms who add incrementally to strip developments that parallel oceanfront access roads, but well-capitalized investors are also creating diversified residential or resort communities on large, virgin blocks of land (such as Hilton Head Island, South Carolina and Amelia Island, Florida). Elsewhere, infilling among already occupied lots is typical.

"More Developed" and "Less Developed" Barriers

Of 282 coastal barriers identified by the U.S. Geological Survey, 171 are listed as partly or wholly developed. These can be divided into two distinct groups: (1) "more developed" barriers, where 50% or more of the land area has been converted to urban or other built-up uses and (2) "less developed" barriers, where some land conversion has occurred but most of the surface remains open (table 3.3).

The distinction between "more developed" and "less developed" coastal barriers has important implications for the reduction of natural hazard losses. "Less developed" barrier communities can choose from a wide range of protection and mitigation alternatives, perhaps including the preservation of existing natural features like dunes, adoption of higher standards for new construction, and growth-management programs to prevent increases in loss potential. "More developed" barriers are likely to confront "damage control" and loss reduction tasks, involving the retrofitting of hazard mitigation measures and the need to provide for redevelopment, either as a consequence of disaster or the normal aging of facilities and infrastructure.

About 120 "less developed" barriers account for 55% of the average of all developed barriers. Most are located in the South Atlantic and Gulf Coast regions (60.5%). Ninety-three of these contain areas somewhat "protected" against devel-

TABLE 3.3

NUMBER AND ACREAGE OF COASTAL BARRIERS
BY DEGREE OF DEVELOPMENT

	New England(a)	Mid-Atlantic(b)	South Atlantic(c)	Florida	Gulf(d)	Total
Number of Barriers						
>50% Developed	15 (33%)	11 (27%)	1 (1%)	24 (30%)	0 (0%)	61 (27%)
<50% Developed	16 (35%)	23 (57%)	26 (36%)	32 (40%)	23 (52%)	110 (39%)
Undeveloped	15 (33%)	6 (15%)	45 (62%)	24 (30%)	21 (48%)	111 (39%)
Total	46	40	72	80	44	282
Acreage of Developed Barriers						
>50% Developed	5898 (48%)	24,421 (61%)	890 (2%)	71,873 (72%)	0 (0%)	103,082 (45%)
<50% Developed	6342 (52%)	15,459 (39%)	43,142 (98%)	27,759 (28%)	32,895 (100%)	125,597 (55%)
Total	12,240	39,880	44,032	99,632	32,895	228,679

SOURCE: H.F. Lins, 1980.

NOTE: Wetlands acreage is included in these data. Although this accentuates regional differences, especially in the Gulf Coast, it does not alter relative contrasts among groups of islands noted above.

opment by CBRA or other mechanisms, but significant numbers of unprotected, "less developed" barriers occur in Florida (6), North Carolina (6), New Jersey (3), and Texas (3). The 51 "more developed" barriers comprise 45% of developed land. All but one are located either in the Northeast (Maine - Maryland) or in Florida. (Sullivans Island, near Charleston, South Carolina, is the exception). Many of the "more developed" Florida barriers contain small parcels of protected land, but few in the Northeast do.

Regional Variations in Development

Using the percentage of acres in "more developed" coastal barriers as an index, it is possible to divide the entire developed barrier population into five distinct regions (table 3.3). The most developed region is Florida (72%), followed by the mid-Atlantic states (61%) and New England (48%). Intensive development of *entire* coastal barriers is negligible in the South Atlantic region (2%) and essentially nonexistent in the Gulf of Mexico, although significant urban centers like Virginia Beach, Galveston, and South Padre Island occupy limited areas.

Different degrees of development imply varying exposure to hazards as well as different mixes of other local problems. Without even considering regional differences in the risk of severe storms, it is therefore unlikely that the same management strategies would be suitable for each region.

Developed Barrier States

The five "development" regions become much more sharply defined at the state level of analysis. Most of the developed acreage (87%), and most of the developed frontage (86%) lie within seven states: Massachusetts, New York, New Jersey, North Carolina, South Carolina, Florida, and Texas (table 3.4). Each of these states, except New Jersey, is typical of the region in which it sits, as far as degree of development is concerned. New Jersey possesses the most developed coastal barriers in the nation.

The pattern of regions with distinctly different levels of development, and key states that contain most of the developed land, has several possible implications for management. It suggests that choices for the management of hazards on developed coastal barriers will be few and constrained in Florida and New Jersey; they will be somewhat more open in the rest of the Northeast; but they will be much more varied and flexible in the South Atlantic and Gulf Coast states. It also suggests that most of the effort to improve hazard management might well be directed towards one or, at most, two states in each region. Finally, different regional mixes of "more developed," "less developed," and "undeveloped" coastal barriers (table 3.3) imply that relationships between the management of protected and unprotected barriers will also vary regionally. Thus, in the South Atlantic region, it should be easier to contain the spillover effects of new development on the existing CBRA system because there are only a relatively few "less developed" barriers set among many undeveloped and protected barriers. Conversely, in New

TABLE 3.4

KEY STATES FOR DEVELOPED COASTAL BARRIER MANAGEMENT

State	Developed Coastal Barrier Acreage	Percent of Barrier Acreage which is Developed	Percent of "More Developed" Barrier Acreage which is Presently Developed
Massachusetts	8,128	3.6%	42.5%
New York	11,578	5.1	56.9
New Jersey	22,719	9.9	72.0
North Carolina	21,625	9.5	0.0
South Carolina	13,081	5.7	6.8
Florida	101,988	44.6	70.5
Texas	19,410	8.5	0.0
Key States Subtotal	198,529	86.8%	52.0%
Other States (11)	30,150	13.2%	24.0%
TOTAL	228,679	100.0%	45.0%

Jersey the few remaining undeveloped coastal barriers (e.g., Brigantine National Wildlife Refuge) are hostage to human activities on the surrounding barriers that are heavily developed. In Florida and New England, there exist relatively even mixes of "more developed," "less developed," and "undeveloped" coastal barriers, often in close association with one another. This suggests very complex management relations between the CBRA system and the developed barriers.

Status of Coastal Barrier Vulnerability Assessments

Some states, such as Florida and North Carolina, have begun comprehensive, coastal barrier land-use classification and mapping tasks. Although Massachusetts has completed a comprehensive inventory of coastal barriers, it offers no information on land use (Smith, 1982). New Jersey, Texas, and Florida have completed detailed vulnerability studies of limited numbers of coastal barriers. Nonetheless, a national vulnerability assessment for developed barriers is presently nonexistent.

The use of developed acreage as an index of hazard vulnerability has several shortcomings. It does not take into account many factors that influence storm and erosion loss potential. These factors include: (1) intensity of development; (2) quality of development; (3) process, rate, and stage of development; (4) degree of seasonal occupancy; (5) effectiveness of hazard preparedness and mitigation measures; and (6) the extent to which undeveloped areas have been negatively altered by human activity.

Hopefully, a more refined analysis of regional vulnerability will be possible as data for these variables become available for all coastal barrier communities. For the present, the blunt instrument used here is perhaps the only one now available that can be applied to all developed barriers as a comparative yardstick.

Conclusion

A developed coastal barrier management program, to be successful, must address the pre-eminent objective of reducing hazard losses in the context of other problems that affect barrier communities. The program should also employ management practices that are compatible with the management of related undeveloped barriers in the CBRA system and elsewhere. This analysis of development on barriers suggest that one strategy for framing a developed barrier management initiative may be to exploit the middle ground between the specificities of CBRA and the broad goals of state coastal zone management programs. It should focus on an organizing principle of multi-state regions with different potential vulnerabilities to hazards and should concentrate on analyzing and responding to the problems of certain key states within those regions.

References

Davidge, R. 1983. "The Development of a Consistent Federal Policy on Coastal Barriers: A Case Study." In *Preventing Coastal Flood Disasters: The Role of the States and Federal Response.* Ed. J. Monday. Special Publication no. 7. Boulder, Colorado: Natural Hazards Research and Applications Information Center, pp. 313-326.

Federal Emergency Management Agency. "National Flood Insurance Program: Claims Information, January 1, 1978 - October 31, 1984." Washington: Unpublished information, 1984.

Lins, H.F. 1980. *Patterns and Trends of Land Use and Land Cover in Atlantic and Gulf Coast Barrier Islands.* Professional Paper 1156. Reston, Virginia: U.S. Geological Survey.

Mitchell, J. K. 1982. "Coastal Zone Management: A Comparative Analysis of National Programs." In *Ocean Yearbook 3.* Ed. E.M. Borgese and N. Ginsburg. Chicago: University of Chicago Press, pp. 258-319.

Mitchell, J. K. 1984. *Hurricane Evacuation in Coastal New Jersey.* Department of Geography Discussion Paper no. 21. New Brunswick, New Jersey: Rutgers University.

New Jersey Department of Environmental Protection, Division of Coastal Resources. 1980. *Shore Protection Master Plan.* Prepared by Dames and Moore. Cranford, New Jersey.

Smith, L. 1982. *Massachusetts Barrier Beaches.* A Report Prepared for the Massachusetts Coastal Zone Management Office by the Provincetown Center for Coastal Studies. Technical Report PCCS 82-1. Provincetown, Massachusetts.

U.S. Department of the Interior. 1980. *Alternative Policies for Protecting Barrier Islands Along the Atlantic and Gulf Coasts of the United States and Draft Environmental Impact Statement.* Washington: Heritage Conservation and Recreation Service.

U.S. Department of the Interior. 1983. *Undeveloped Coastal Barriers: Final Environmental Impact Statement. Washington: Coastal Barriers Task Force.*

THE POLITICAL GEOGRAPHY OF DEVELOPED COASTAL BARRIERS

Rutherford H. Platt and Keane Callahan

University of Massachusetts

Introduction

In their natural state, coastal barriers possess well-recognized geomorphic features: offshore bars, beaches, dunes, uplands or fastlands, bay shores, and salt marshes. These features vary in composition, appearance, and extent according to local conditions of wind, tide, beach material, and storm history. But the same natural laws governing barrier dynamics apply equally from Maine to Texas and wherever else such features are found.

Human settlement of coastal barriers introduces complex new variables, including the laws, policies, and actions of various public authorities, as well as individual decisions by private landowners, investors, and visitors to coastal barrier communities. These in turn reflect objectives that are frequently in mutual conflict, such as preservation of coastal amenities, private profit, public tax revenue, avoidance of natural hazard losses, and public recreation. A highly variable pattern of coastal development thus results from the interplay of both natural and human forces.

In 1968, Burton, Kates, and Snead identified four types of human occupance of the shoreline of "Megalopolis": (1) village, (2) urban, (3), summer, and (4) empty places. Today we find these classifications blurred as former summer colonies become winterized for year-round use, high-rise condominiums supplant the "village" atmosphere of former fishing communities like Ocean City, Maryland, and major development corporations reshape the landscape of a Hilton Head or Amelia Island.

Fragmentation of Authority

The management of these disparate barrier communities is the shared responsibility of public and private authorities. Such authority is fragmented both "vertically" and "horizontally." In the vertical dimension, authority to affect the use of land on developed coastal barriers, like other land resources in the United States, is divided among a pyramid of authorities. These include the federal, state, county, and local governments, and relevant special districts, as well as individual property owners (either public or private). Public and private authority is fragmented horizontally among adjoining units of ownership and public authority. (Platt et al., 1980).

Management authority over coastal barriers is thus pluralistic. No single unit of authority has exclusive jurisdiction over any portion of the coastal barrier chain, with the possible exception of the federal government with respect to its own holdings. The picture is further complicated by the fact that each level of public authority is internally fragmented along functional or administrative lines, e.g., the National Flood Insurance Program and the federal Coastal Zone Management Program (Platt, 1978). Similarly, at the state level a Department of Natural Resources may conflict with a Department of Economic Development regarding the relative merits of preserving *versus* developing a given coastal site. Competition between mutually conflicting land uses, public and private objectives, and even between public objectives is endemic to coastal zone management generally, and barrier island management in particular.

Federal and state roles in managing barrier resources have been reviewed by others. This paper is concerned with the effects of political fragmentation upon local management of developed coastal barriers. Specifically, our study raised several interrelated questions: (1) What are the immediate and long-term impacts on developed coastal barriers which result from fragmentation of political authority? (2) What are the planning capabilities of local barrier jurisdictions? (3) Do responses to the first two questions differ according to the type of barrier community considered?

Local Jurisdictions on Developed Coastal Barriers

The dominance of local government in American land-use planning was noted by the British observer John Delafons (1969, p. 9):

> [In the United States] there are no effective planning authorities covering more than one local government unit; each municipality is its own planning agency, and the power of land-use control is one activity which is never relinquished to another authority. Inevitably it is made to serve essentially local interests, and by very general admission, private interests are more likely to be observed than any conscious public objective.

This appraisal is perhaps less applicable to the coastal zone in the 1980s than to metropolitan areas generally in the 1960s. Federal and state legislation of the past fifteen years has whittled away at local autonomy. Nevertheless, local governments continue to exercise substantial powers under state grants of authority or

TABLE 4.1

COASTAL BARRIERS -- SUBSTATE AUTHORITIES

I. Incorporated Municipalities
   A. Freestanding
      1. Entire Barrier
      2. Segment of Barrier
   B. Appendant to Mainland Municipality
      1. New England Town or City
      2. Other Mainland Municipalities

II. Unincorporated Places -- County Jurisdiction

III. "Quasi-Governments"
   A. Homeowners Associations
   B. Development Corporations

IV. Special Districts (e.g., sewer and water districts; soil and water conservation districts, etc.)

home rule provisions which directly affect the quality and quantity of new development on coastal barriers. Such powers typically include the functions of land-use planning, zoning, and subdivision regulation, the provision of public facilities such as streets, sewers, water lines, and public beaches, and responsibility for disaster planning, including hurricane evacuation. Local governments thus exert powerful influence over the form, location and timing of development within their jurisdictions, but have little voice as to what happens in neighboring jurisdictions. It is therefore essential to recognize the legal nature and geographic extent of coastal units of local government authority (table 4.1).

The political geography of local governments is highly variable. A major difference among states involves the extent of municipal incorporation. The coastline of the New England states is entirely subject to municipal governance by incorporated cities or towns; there is no unincorporated land. It follows that local towns and cities in New England include substantial areas of undeveloped land, although this is increasingly rare on coastal barriers. By contrast, municipal incorporation is the exception rather than the rule along the Southeast Atlantic and Gulf shorelines. In those areas, municipalities are generally confined to a given built-up area. Much development occurs in unincorporated areas under the jurisdiction of county governments.

Another variable to consider is the geographic extent of local government units. A dichotomy is evident between communities which are "freestanding," i.e., limited to the coastal barrier itself, *versus* those which are "appendant" to a larger mainland municipality (Platt, 1985). In New England, the latter situation prevails since all barrier beaches are part of larger town or city units. Elsewhere, unincorporated communities are likewise appendant to mainland counties. Individual municipalities outside New England may be either freestanding or appendant. Freestanding municipalities range from tiny vacation or fishing enclaves to major cities such as Atlantic City, Miami Beach, and Galveston. Appendant coastal barriers are contained within larger municipal units ranging in size from, e.g., Clearwater, Florida to New York City (which contains Rockaway Beach and Coney Island).

An additional complication is that many built-up areas do not fall neatly within a single unit of local or county jurisdiction. Much of the New Jersey barrier beach shoreline, for example, is continuously built up, yet is divided into discrete municipal units like beads on a string. Conversely, a given municipal unit, especially in New England, may contain more than one developed coastal barrier settlement. A single island may have several types of governing authorities side by side. For example, Fire Island, New York is divided among town level units, incorporated villages, and private associations having quasi-public authority over their own territories through deed restrictions. All of these are intermingled with state and federal holdings. Amelia Island, Florida is divided between a public municipality, a private corporation, and unincorporated land under county jurisdiction.

Given the extreme variability in the types of local government units that administer developed coastal barriers, is it conceivable that experience in one location could be useful anywhere else? We believe that it is. But essential to the transfer of experience from one setting to another is recognition of the political geographic context in whch such experience has occurred.

Inventory of Developed Coastal Barrier Jurisdictions

In the absence of any such list in the literature, we compiled an inventory of local jurisdictions that encompass developed coastal barrier settlements. Our sources of data included state Coastal Zone Management (CZM) reports, USGS topographic maps, and U.S. Census of Governments reports. Our list includes 234 distinct political units.

This inventory of developed coastal barrier jurisdictions is subject to certain qualifications. Some municipalities may be too recent to have been recorded in published sources. Others may be so small that they have escaped notice. Many of the jurisdictions listed are not entirely developed and indeed contain units of the coastal barrier resources system within their borders in some cases. Likewise, many built-up areas are adjacent to or surrounded by public holdings in the form of defense installations, federal and state parks, and other public lands. The geographic boundaries of county and municipal governments may encompass portions of such public lands, although the local unit has no legal control over the use of federal or state land. In other words, local jurisdictions may be geographically broader than their effective area of control.

Survey Findings

This inventory provided the basis for a survey of local planning issues and practices in coastal barrier communities. To generate a representative sample for surveying, the barrier communities were classified under nine headings, according to demographic and geographic criteria. A stratified random sample of communities was then selected from each class, resulting in a total sample size of 52.

A mail survey was designed to gather the following information: (1) community characteristics, such as community incorporation, economic activity, and population trends; (2) political jurisdictions; (3) development patterns; (4) tourist facilities and impacts; (5) planning concerns; and (6) growth management efforts. A total of 36 questionnaires were returned, yielding a response rate of 69%. A number of statistical tests were conducted to determine the relationship between political fragmentation and coastal barrier planning and management problems. All relationships were tested at the .05 level of significance. Phi coefficients were noted to indicate the strength of the relationships.

Community Characteristics

Most of the coastal barrier communities (78.4%) are incorporated and self-governing. Approximately 30% of these were incorporated before 1900 and nearly 38% between 1900 and 1960. Only about 11% have been incorporated since 1960. This leaves 21.6% of the sample in unincorporated status under county jurisdiction. The survey revealed regional variation as to the date of incorporation. Compare, for instance, the Northeast Atlantic (Maine-Maryland) and Southeast Atlantic (Virginia-Florida Atlantic Coast) with respect to the period of incorporation. Nearly 43% of Gulf Coast sample communities were incorporated between 1900 and 1960, with no further incorporations since the latter date (fig. 4.1).

As might be expected, tourism or resort development and associated commercial activities serving the tourist population are the primary economic activities for the majority of the sample communities (fig. 4.2). Close to 61% of the survey respondents indicated that tourism is pivotal to their economic viability, even though most of this activity is seasonal. Other economic components, such as second home construction, citrus production, and educational institutions (e.g., universities and research facilities) are also important to developed coastal barriers. Fishing occupies a very modest position as a primary economic activity.

The demographic characteristics of developed coastal barriers are diverse. Sixty-seven percent of the survey communities have permanent populations of fewer than 10,000; 44% have populations under 2,500. The mean permanent, year-round population of the survey communities is approximately 15,800. Peak seasonal populations, including permanent and summer residents and day or overnight visitors, approaches a mean of 53,000, or an increase of 231% over the mean permanent population. Further analysis of the survey data indicated that the larger seasonal influx of population occurs primarily in smaller communities. For example, survey communities with populations under 10,000, on an average, experienced a 100-fold increase from year-round to seasonal populations. In

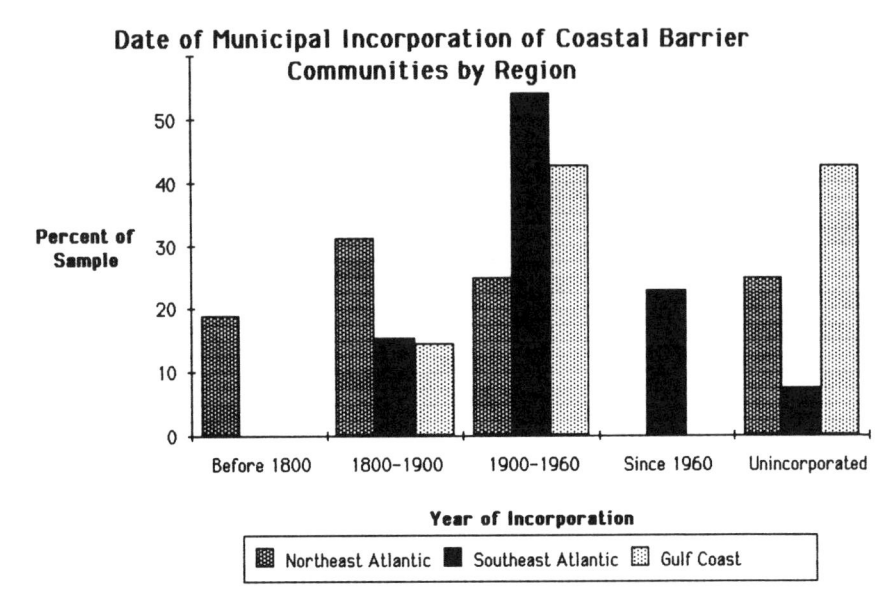

Date of Municipal Incorporation of Coastal Barrier Communities by Region

Figure 4.1

Primary Economic Activity of Sample Coastal Barrier Communities

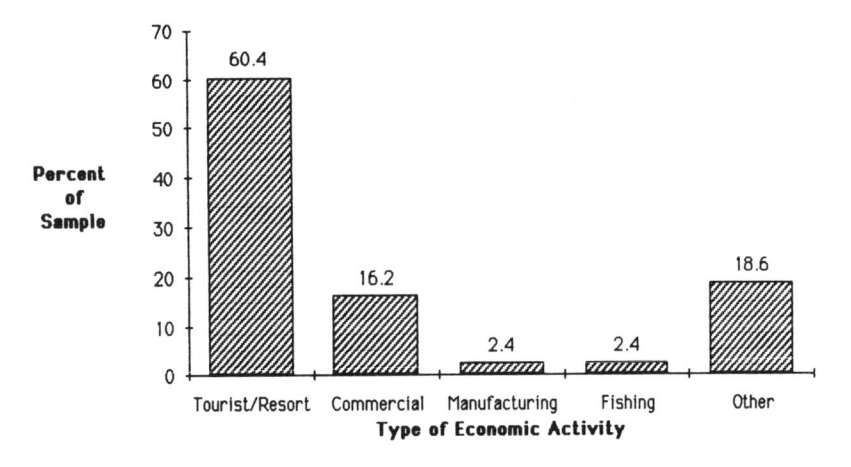

Figure 4.2

contrast, survey communities with populations greater than 10,000 experienced only a 5-fold increase in population.

Development Trends

Lins (1980) analyzed recent trends in development on Atlantic and Gulf Coast barrier islands. Of the 282 coastal barriers surveyed, Lins found many to be extensively developed, such as Atlantic City, New Jersey and Wrightsville Beach, North Carolina. He estimated that land used for urban development had risen from 5.5% to 14% of total coastal barrier acreage between 1945 and 1975, an increase of 153%, and most of this development had occurred in wetlands and forested upland areas.

Our survey communities reported where development was occurring most rapidly on specific physical features illustrated in the following table:

Coastal Barrier Feature	Percentage of Respondents
Active Beach	13.5%
Dune	35.3%
Fastland (Upland)	55.8%
Wetland	23.5%

The comparatively low figure for wetland development may reflect the restraining influence of more recent federal, state, and local wetland protection laws. The proportions of communities reporting development of beaches and dunes, despite the well-documented hazards of doing so, suggests the need for more stringent public controls over these resources.

Reporting on a study by Sheaffer and Roland (1981), Miller stated that there is a propensity for coastal barrier communities to develop to the greatest extent possible, pushing existing infrastructure to its limits (Miller, 1981, p. 6). Our survey confirmed this trend. Fifty-eight percent of the survey respondents indicated that their community was more than 50% developed, with over one-third reporting that they are 75-100% developed (fig. 4.3). Development of this intensity presumably extends into coastal hazard areas subject to wave and wind damage, storm surge, and overwash.

Horizontal Fragmentation

Survey respondents were asked to specify the types of governmental units, if any, that exercise jurisdiction over their community. The situation can become confusing since all coastal barriers are technically subject to state and federal authority and most are subject to county jurisdiction as well. Our purpose was to determine whether portions of the respondent's barrier are occupied by other municipalities, by unincorporated areas under county jurisdiction, or by facilities *under* state or federal control. Respondents indicated the presence of other governmental jurisdictions as follows:

Types of Other Political Jurisdictions	Percentage of Respondents
Other municipalities	53.2%
County (unincorporated areas)	71.8%
State facilities	46.6%
Federal facilities	41.9%

Thus, 53.2% of the respondents reported that at least one other municipality shares the same coastal barrier. For example, Island Beach in New Jersey is occupied by five separate, incorporated communities.

Counties have jurisdiction over unincorporated areas, and such counties as Sarasota County, Florida also have jurisdiction over erosion control projects and the formulation and implementation of beach management plans.

Our survey indicated that state and federal governments play active roles in the planning and management of coastal barriers within their jurisdictions. Many states have established state parks and seashores on barrier beaches, protecting limited areas from further development pressure. The federal government also has acquired a number of coastal barriers and designated them as national seashores. Finally, the federal government has responsibility or jurisdiction over Army Corps of Engineers beach nourishment projects and the Intracoastal Waterway.

Political fragmentation varies regionally as well. Many coastal barriers in the North Atlantic and Gulf Coast regions are dominated by municipalities. For example, Sand Key on Florida's Gulf Coast is shared by nine incorporated

Extent of Development in Coastal Barrier Communities

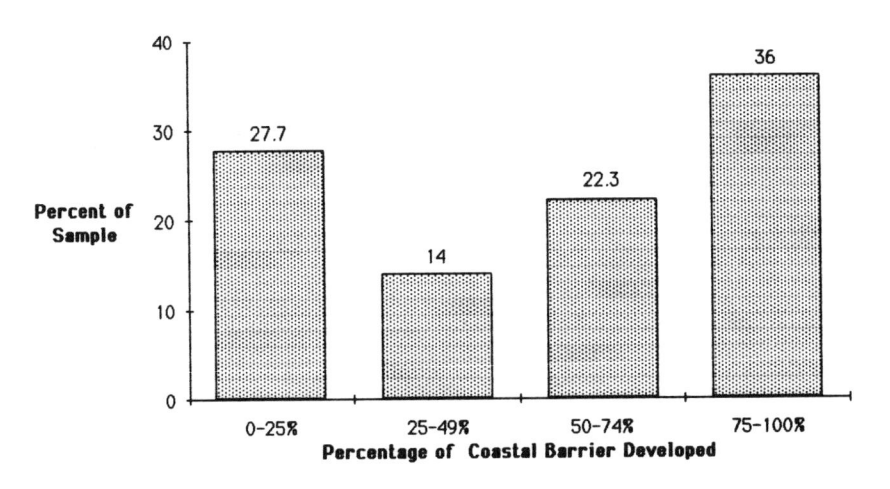

Figure 4.3

municipal jurisdictions. The New Jersey coast shows similar characteristics. County jurisdictions are prevalent in all three regions, particularly along the Southeast Atlantic and Gulf coasts. State facilities are evenly distributed throughout all regions. Federal facilities are most prominent in the Northeast Atlantic region.

Impact of Fragmentation on Planning

Twelve survey communities responded that at least three or more jurisdictions had some authority over their island community. Thus, more than 36 separate, often conflicting or incompatible jurisdictions are responsible for the management of those coastal barriers. Eight communities or 22% of the sample reported a total of five separate governmental or quasi-governmental jurisdictions (e.g., special districts and homeowners' associations overlapping their local jurisdiction).

Statistical analysis of the survey data revealed that there is a significant relationship between multiple political jurisdictions and the extent of planning problems on developed coastal barriers (chi-square = 11.7, 5 df, $p < .05$, Phi = .247). The Pearson Product-moment Correlation test was also performed, and it indicated that a fairly strong correlation exists between the two variables ($r = .403$, 34 df, $P < .05$). Thus, our survey revealed that a number of planning problems do arise from or are at least associated with political fragmentation. For example, 64% of the survey communities responded that conflict among planning goals, policies, and laws was a primary concern caused by multiple jurisdictions; 52% noted the lack of intergovernmental cooperation and communication as a problem; uneven or inconsistent local management control was also cited by 42% of the respondents. Other effects of political fragmentaton were conflicting developmental patterns (34%), restricted beach access and use (32%), and loss of critical environmental areas (28%) (fig. 4.4).

Problems Caused by Political Fragmentation

Planning Problems	Percent of Sample
Loss of natural areas	28.2
Restricted beach access	32.3
Conflicting development	34.4
Lack of cooperation	51.5
Inconsistent local control	42
Conflicting planning goals	63.6

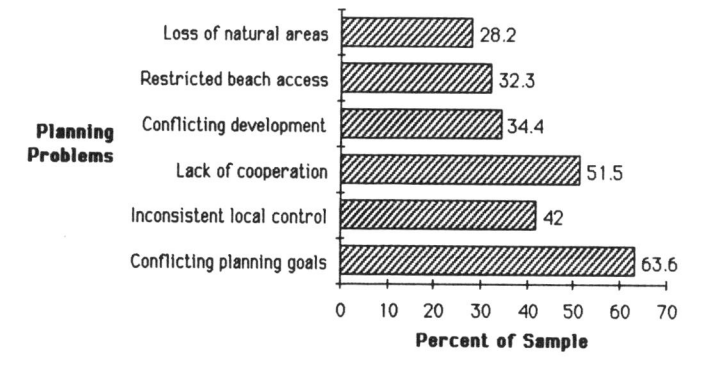

Figure 4.4

Conclusion

All coastal barriers are governed by multiple units of political authority that are jointly responsible for the management and planning of barrier resources. In some instances, this state of fragmentation causes inconsistent or conflicting planning goals at the local level. For example, many coastal communities are economically dependent on the tourist/beach resort industry, as demonstrated by our survey. The development or encouragement of this industry can create negative environmental impacts such as the filling of wetlands or the destruction of important dune systems. However, states and the federal government have recognized the importance of these ecosystems as statewide and national resources requiring sound public management. Subsequently, legislation has been enacted to protect significant natural areas. These actions often conflict with local objectives of community and economic development. The opposite is also true: numerous barrier communities have more stringent environmental protection laws than either the state or federal government. The federal government may promote the development of a major energy or navigational facility that directly competes with local planning efforts to conserve coastal environments.

This conflict has a number of implications for coastal barrier communities. First, tourism is critical to the economies of the majority of barrier communities and it must be balanced by wise development and conservation of resources in order to maintain the natural amenities that promote recreational use and tourism. Second, multiple political jurisdictions are prevalent on developed coastal barriers and are capable of influencing local planning goals or decisions. Barrier communities therefore need to establish a planning process that will effectively manage coastal resources within this multi-jurisdictional framework. Intergovernmental communication and cooperation is thus imperative in order to formulate innovative plans to provide for public recreational needs and to guide growth on coastal barriers.

References

Burton, I., R.W. Kates, and R.E. Snead. 1968. *The Human Ecology of Coastal Flood Hazard in Megalopolis*. Research Paper no. 115. Chicago: University of Chicago Department of Geography.

Delafons, J. 1969. *Land-use Controls in the United States* (2nd ed.). Cambridge: The M.I.T. Press.

Lins, H.F., Jr. 1980. *Patterns and Trends of Land Use and Land Cover on Atlantic and Gulf Coast Barrier Islands*. Geological Survey Professional Paper 1156. Washington: U.S. Government Printing Office.

Miller, H.C. 1981. "The Barrier Islands: A Gamble With Time and Nature." *Environment* 23: 6-10, 36-42.

Platt, R.H. 1978. "Coastal Hazards and National Policy: A Jury-Rig Approach." *Journal of the American Institute of Planners* 44: 170-180.

Platt, R.H. 1985. "Congress and the Coast." *Environment* 27: 12-17, 34-40.

Platt, R., et al. 1980. *Intergovernmental Management of Floodplains*. Hazard Monograph no. 42. Boulder: University of Colorado Institute of Behavioral Science.

Sheaffer & Roland, Inc. *Barrier Island Development Near Four National Seashores*. Chicago: 1981 (mimeo).

POPULATION CHANGES IN COASTAL JURISDICTIONS WITH BARRIER BEACHES: 1960-1980

Niels West

University of Rhode Island

Introduction

Limits on federal expenditures imposed by the Coastal Barrier Resources Act of 1982 are based on a distinction between "developed" and "undeveloped" coastal barriers. The developed areas were arbitrarily defined as having at least one dwelling unit per five acres. The significance of this criterion was not empirically tested prior to the passage of the bill. It remains to be seen whether this definition will delineate a meaningful geographic unit for management purposes. In any event, the concept of "developed barriers" defies easy analysis in terms of demography.

Few studies have specifically addressed demographic changes in the coastal zone or population distribution on coastal barriers. This paper attempts to analyze coastal population changes between 1960 and 1980 at both the county and sub-county levels and discusses why demographic research is especially needed in this area.

The Bureau of Census (1982, p. 7) has identified 611 counties and independent cities which are "entirely or substantially within 50 miles of U.S. coastal shorelines." Units within 50 miles of the Atlantic and Gulf coastlines increased in population from 34.1 million in 1940 to 63.3 million in 1980, an increase of 85% as compared with 70% for the nation as a whole. Gulf Coast counties, considered alone, grew by 200%.

In 1976, the author undertook a coastal county analysis incorporating demographic changes between 1950 and 1974. That study suggested that growth in coastal metropolitan counties will no longer continue at rates comparable to those of the immediate past. The reasons may be either economic or environmental. It is entirely possible that counties in coastal SMSAs are becoming saturated under

TABLE 5.1

REGIONAL POPULATION DENSITIES IN COASTAL COUNTIES
WITH COASTAL BARRIERS 1960-1980

(pop. per square mile)

Region (No. of Counties)	1960	1970	Percent Change 1960-1970	1980	Percent Change 1970-1980
New England (11)	490	540	10.2	571	5.7
North Ctr. Atlantic (9)	736	940	27.7	1233	31.2
South Ctr. Atlantic (11)	108	144	33.3	178	23.6
South Atlantic (11)	175	263	50.3	381	44.9
Gulf (16)	146	190	23.2	255	34.2

SOURCE: U.S. Bureau of the Census, 1982.

existing zoning ordinances. New developments, especially large subdivisions, may take place in less developed, inland counties, where land is often more readily available, less expensive, and better able to sustain new development.

Numerous studies have addressed the attraction of water for residential, recreational, commercial, and industrial activities. Yet none have applied existing models or developed new ones for the demographic problems encountered in the coastal zone. In order to make small area population projections, density statistics are usually required. Unfortunately, this information is not routinely available for jurisdictions below the county or city level.

Analysis

The coastal barriers studied here extend from New England to the Mexican border and include communities from the industrial Northeast to the Sunbelt. This chain of barriers has been divided into the following five regions: (1) New England (Maine, New Hampshire, Massachusetts, and Rhode Island); (2) North Central Atlantic (New York, New Jersey, Delaware, and Maryland); (3) South Central Atlantic (Virginia through the Carolinas to Georgia); (4) South Atlantic (Florida's Atlantic Coast); and (5) Gulf (west coast of Florida, Alabama, Mississippi, Louisiana, and Texas).

Table 5.1 summarizes the demographic characteristics of the coastal regions for the period 1960 to 1980, based on density estimates for counties containing

POPULATION CHANGES

TABLE 5.2

REGIONAL POPULATION CHANGES IN 156 LOCAL JURISDICTIONS
CONTAINING COASTAL BARRIERS: 1960-1980

Region	1960	1970	Percent Change 1960-1970	1980	Percent Change 1970-1980
New England	251,915	362,133	43.7	464,823	28.0
North Ctr. Atlantic	1,993,414	2,243,803	12.6	2,218,247	- 1.1
South Ctr. Atlantic	35,506	206,320	484.3	330,761	60.3
South Atlantic	359,872	653,294	81.5	831,587	27.1
Gulf	1,043,177	1,292,571	23.9	1,631,137	26.2

SOURCE: U.S. Bureau of the Census, 1982.

coastal barriers. Spatio-temporal population density changes in the five regions appear to substantiate several of the urban-rural changes that occurred nationwide during this period. The central city populations grew rapidly during the 1960s, then stabilized during the 1970 decade. Some cities in the Northeast lost population during this period.

The North Central Atlantic coastal region is by far the most densely populated sector of the study area. At the county level, this region experienced robust growth during both decades, in contrast to the New England area which grew only moderately. The demographic changes in the South Central Atlantic, South Atlantic, and Gulf regions are particularly important indicators of future trends, especially on the sub-county level. The South Atlantic region experienced very dramatic growth from 1960 to 1970 and tapered off only slightly during the 1970s. This pattern was also evident in the South Central Atlantic and Gulf regions, but to a lesser extent.

The ensuing analysis is based on 156 local municipalities which include coastal barriers as part of their jurisdiction. Since the municipal boundaries rarely if ever coincide with the boundaries of the natural systems, the analysis is based on those municipalities whose territories include all or part of a coastal barrier. Coastal barriers which are located within a mainland municipality are sometimes referred to as "appendant" jurisdictions. In others, a single coastal barrier may be divided among several municipalities. Communities entirely situated on a barrier are termed "freestanding" (Platt, 1985).

The data in table 5.2 consists of tallies from incorporated coastal jurisdictions below the county level, including towns, villages, boroughs, and cities. These data tend to verify trends that have been known to exist in the country as a whole. The physical amenities associated with many coastal communities have made them particularly susceptible to rapid, seasonal and permanent residential development. The latter phenomenon is particularly troublesome for those interested in estimating and projecting demographic changes, since many seasonal residences are actually being used for year-round housing.

In New England during the 1960s, coastal communities grew at an annual rate in excess of 4%. Although this growth continued during the 1970s, it has tapered off somewhat during the subsequent decade. Coastal jurisdictions in the South Central Atlantic region also show this trend, no doubt influenced by the availability of developable land, rapid economic growth which occurred throughout the Southeast during this period, and the attractiveness of the region's coastal environment.

The very densely populated North Central Atlantic was the only region that did not exhibit this growth. Many of the coastal jurisdictions in this region already were substantially developed by the 1960s, leaving little space for further infilling. Other reasons may account for the demographic trends that are evident in these two tables. These may include the size of the statistical jurisdictions. The geographically smaller towns and counties in the northeast are a function of their historical settlement patterns. Municipal boundaries were often established at the time of initial settlement which, combined with a very strong home rule tradition, made annexation of land by one community both culturally and politically difficult. Urbanization in other areas has often been accompanied by annexation of adjoining unincorporated areas. The process of annexation complicated intertemporal analysis in the South Central and Gulf regions, since some of their reported population change may be due to annexation of new areas. Also, the smaller communities of New England and the North and South Central Atlantic sub-regions are more likely to reach a "build out" situation sooner than larger jurisdictions, or those which have changed size through annexation.

The trends evident on the county level are not replicated on the local level. Mention was made earlier that counties included in the North Central Atlantic region grew rapidly during the two census periods (table 5.1), yet when analyzing the demographic trends on the local level, a different picture emerges. Coastal communities in this region grew only moderately during the 1960s and, in fact, lost population during the 1970s. Similarly, the dramatic population growth on the county level which took place in the South Atlantic, while strong, is exceeded substantially by the rate in the South Central Atlantic during both periods (table 5.2).

The following analysis is based on the regional patterns of those jurisdictions that lost population or experienced sluggish growth, here defined as communities that grew less than 50% during the decade. Moderate growth ranges between 50.1 and 100% during the decade and rapid growth is defined as growth exceeding 100% during the decade. The information has been summarized in tables 5.3 and 5.4. Several observations can be made, based on the information displayed here, which tends to substantiate the discussions above. First, the demographic changes appear to have both a temporal and a spatial content. During the 1960s, all five regions sustained population losses in certain places.

POPULATION CHANGES

TABLE 5.3

COASTAL MUNICIPAL POPULATION LOSSES AND GAINS BY REGION 1960-1970

Region		Losses	0-50% Gain	50.1-100% Gain	More than 100% Gain
New Eng.	Raw Score	4	43	9	5
	Row Percent	6.6	70.5	14.8	8.2
	Col. Percent	14.8	53.8	40.9	18.5
North Ctr. Atl.	Raw Score	2	11	4	1
	Row Percent	11.1	61.1	22.2	5.5
	Col. Percent	7.4	13.8	18.2	3.7
South Ctr. Atl.	Raw Score	5	8	1	6
	Row Percent	25.0	40.0	5.0	30.0
	Col. Percent	18.3	10.0	4.6	22.2
South	Raw Score	6	7	4	12
	Row Percent	20.7	24.1	13.8	41.4
	Col. Percent	22.2	8.8	18.2	44.4
Gulf	Raw Score	10	11	4	3
	Row Percent	35.7	39.3	14.3	10.7
	Col. Percent	37.0	13.8	18.2	11.1

SOURCE: U.S. Bureau of the Census, 1982.

Furthermore, with the exception of the New England and the Gulf regions, the number of local jurisdictions that lost population in the three central regions (North Central, South Central, and South Atlantic) was approximately what one would expect under random conditions. Most of the municipalities that experienced growth in excess of 100% were located in the South Atlantic and South Central Atlantic regions.

The spatial trends that became evident during the 1960s appear to have continued through the 1970s. More than 73% of the communities that lost population between 1970 and 1980 were located on the perimeter of the South Central Atlantic growth region. Nearly half of the municipalities in this region at least doubled in size during this decade. The importance of this region as a demographic growth node is supported by the fact that during this period 35% of this region's municipalities at least doubled in population. The South Central Atlantic also included seven out of the 15 municipalities that doubled in size during this decade in all regions.

TABLE 5.4

COASTAL MUNICIPAL POPULATION LOSSES AND GAINS
BY REGION: 1970-1980

Region		Losses	0-50% Gain	50.1-100% Gain	More than 100% Gain
New Eng.	Raw Score	15	30	15	1
	Row Percent	24.6	49.2	24.6	1.6
	Col. Percent	44.1	40.0	46.9	6.7
North Ctr. Atl.	Raw Score	3	12	2	1
	Row Percent	16.7	66.7	11.1	5.6
	Col. Percent	8.8	16.0	6.3	6.7
South Ctr. Atl.	Raw Score	2	6	5	7
	Row Percent	10.0	30.0	25	35
	Col. Percent	5.8	8.0	15.6	46.7
South	Raw Score	4	15	5	5
	Row Percent	13.8	51.8	17.4	17.4
	Col. Percent	11.8	20.0	15.6	33.3
Gulf	Raw Score	10	12	5	1
	Row Percent	35.7	42.9	17.9	3.6
	Col. Percent	29.4	16.0	15.6	6.7

SOURCE: U.S. Bureau of the Census, 1982.

Findings

The availability of demographic information is related to the scale of the political jurisdiction. The smaller the geographic unit, the less information is generally available. By using different data sets, this study has attempted to assess the demographic trends in coastal communities containing developed coastal barriers. Even though the political jurisdictions from which the data have been obtained rarely correspond to the physical boundaries of the coastal barrier, several inferences can be drawn from the analysis:

1. Coastal demographic developments appear to follow national trends, with New England showing the slowest relative increase in population during the 1960s

2. Absolute growth in population during the 1970s was centered in the three southern subregions (South Central Atlantic, South Atlantic, and Gulf) with the South Central Atlantic emerging as the demographic growth node
3. Urbanization of the coastal barriers appears to be continuing at a rate faster than that of inland locations
4. Coastal urban population development appears to mirror the demography of central cities in the country as a whole
5. The demography of communities with coastal barriers may be approaching the point where future population changes will begin to level off

Conclusions and Recommendations

Environmental resource management decisions in the U.S. have commonly been made under two sets of constraints. First, many federal and state regulations have been issued without adequate social and environmental information. This has been a particularly troublesome problem in the context of environmental planning and management where adequate baseline data often are not available and/or may not become available within a politically viable timeframe. Second, many decisions are crisis-influenced. Decisions made under these conditions are usually made without adequate time to collect necessary information, undertake required analysis, or formulate and implement viable management plans. As a result, such decisions are, at their best, inefficient and, at their worst, totally unworkable. The classification of "developed" and "undeveloped" coastal barriers appears to have been constrained by at least the first of the two factors.

The first recommendation addresses the absence of suitable data and information sources for threatened environments (including barrier islands and beaches). Data should be collected regularly, not only pertaining to barrier beaches *per se*, but to coastal jurisdictions in general. Furthermore, such information should include both social and physical parameters. In short, the coastal barriers should be viewed as a socio-environmental system where social and physical activities and processes constantly interact. Finally, this information should be collected at the federal level and in accordance with some universally agreed upon standards and procedures. This will help insure continuity and comparability on both a spatial and a temporal basis. The Office of Ocean and Coastal Resource Management of NOAA should be the lead agency responsible for collecting and disseminating this information.

The second recommendation urges the academic community to begin to test, modify, and, where necessary, develop new demographic models that have specific applicability to coastal areas. The uniqueness of the coastal region, in terms of physical and other characteristics, suggests that such a goal will be difficult to accomplish. The intensive use of most developed coastal barriers underscores the need for these models so that better planning and management can be achieved.

Such modeling efforts might include variations on the "gravity" and "suitability" models. While the gravity or "distance decay" model has been used to describe and estimate land uses (Chisholm, 1962) and a host of socioeconomic activities between neighboring and competing urban (economic) centers, the model has also measured the attractiveness and thus the usage of a given recreational or residential facility (Clawson, 1959; Clawson and Knetsch, 1966; McAllister, 1975). The underlying assumption common to all the variations of this model is the declining intensity of land usage with increasing distance from a city (demand) center. In order for this model to be applicable in a coastal context, the environmental attractiveness concept would have to be expanded in two dimensions. In addition to the traditional urban/rural axis, where population densities are hypothesized to decline with increasing distance from the urban (demand) center, the modified coastal gravity model would have to incorporate the attractiveness of the coastal region itself, which extends inland from the shoreline. An oceanfront property is certainly more attractive than a similar lot without access to the water. Although difficult to quantify, some attempts have been made to measure such increases in value (Gillard, 1981). Furthermore, the increased attractiveness has often resulted in very high densities in areas adjacent to the shoreline. The very densely populated barrier beaches of New Jersey and Florida are both examples of such developments. Either area may be useful in testing the validity of the modified (two-dimensional) gravity model.

Other research efforts have been undertaken which may have applicability to estimating and projecting populations in the coastal zone. The "suitability model" is based on the socio-environmental carrying capacity of the region, past growth rates, and the extent to which undeveloped (raw) land is still available for residential development. This model is an extension of the work of Leopold et al. (n.d.) in assessing environmental impacts in the wilderness areas, except that it incorporates cultural constraints and opportunities in addition to those inherent in the physical environment.

Variations of the gravity and environmental suitability models may be useful starting points for improving current population projection techniques in the coastal zone, with particular emphasis on the development of coastal barriers.

Endnote

The raw data identifying the individual coastal barriers, their local jurisdictions, county, state, and population changes is available from the author. Space limitations prevented their inclusion in the present manuscript.

References

Chisholm, M. 1962. *Rural Settlement and Land Use*. London: Hutchinson University Library.
Clawson, M. 1959. "Methods of Measuring the Demand for a Value of Outdoor Recreation." Resources for the Future (RFF) Reprint #10.
Clawson, M., and J. Knetsch. 1966. *The Economics of Outdoor Recreation*. Baltimore: Resources for the Future and the Johns Hopkins University Press.
Gillard, Q. 1981. "The Effect of Environmental Amenities on Housing Values: The Example of a View Lot." *Professional Geographer* 33(2): 216-220.
Leopold, L.B., et al. (n.d.) "A Procedure for Evaluating Environmental Impact." U.S.G.S. Circular #645.
McAllister, M. 1975. "Planning an Urban Recreation System: A Systems Approach." *Natural Resource Journal* 15(3): 567-580.
Platt, R.H. 1985. "Congress and the Coast." *Environment* 27(6): 12-17, 34-40.
U.S. Bureau of the Census. 1982. *Statistical Abstract of the United States 1982-1983*. Washington: U.S. Government Printing Office.
U.S. Bureau of the Census. 1982. *Census of Population*. Washington: U.S. Census Bureau, Dept. of Commerce. (Several volumes.)
West, N. 1976. "Coastal Demographic Changes in the U.S. 1950-1974." *Ocean 80 Proceedings*.
World Bank. 1984. *World Development Report, 1984*. New York: Oxford University Press.

SHORELINE CHANGES ON DEVELOPED COASTAL BARRIERS

Karl F. Nordstrom

Rutgers - The State University
of New Jersey

The Issue

A major source of debate among persons charged with managing the coastal zone has been the issue of the desirability of building and protecting fixed structures on dynamic coastal barriers. Opinions on this issue appear to be polarized, with coastal geomorphologists, ecologists, and environmental groups urging that the correct management strategy is to promote a naturally functioning barrier beach system (Pilkey, 1981). Coastal engineers, land developers, and builders, however, argue that such dynamic environments can be developed without unwarranted hazard risks or environmental losses (O'Brien and Johnson, 1980; Parker, 1980; Vaccaro, 1981).

Geomorphic data supporting either side of the argument on suitability of development are lacking. There are numerous studies identifying the effects of waves and winds on buildings, but studies of the effects of buildings on the geomorphic processes of the natural system are virtually nonexistent (Nordstrom and McCluskey, 1985). The effects of individual protection structures such as groins and revetments have been identified in design studies, but documentation of actual structures is lacking. There are technical studies of the effects of engineering structures on regional shoreline change, but these studies are too few, given the magnitude of the problem and the pressing need for data.

One purpose of this paper is to identify the problems geomorphologists encounter in obtaining these critical data. The basic problem does not lie in demonstrating that there is a change in coastal barriers under conditions of development; rather, it lies in demonstrating that an observed change can be directly related to a specific series of human adjustments, and that the change is adverse. Difficulties in interpreting the causes of observed changes have led to misinterpretation of the effects of development or to the conclusion that all fixed buildings or shore protection efforts are incompatible with a coastal barrier location.

This paper draws upon experience with development on New Jersey's barrier shoreline. It is assumed that the New Jersey shoreline will continue to be used intensively in the future. In this context, the geomorphologist must first determine whether and to what extent human uses cause significant adverse effects on natural systems. Then he or she must work with planners and public officials to develop strategies that allow for short-term development while retaining the components of the natural system which cause these locations to be so highly valued.

At a minimum, system preservation means retention of a beach in urban areas and retention of natural habitats in remaining undeveloped areas. These goals have been expressed in federal legislation on beach nourishment (Adams, 1981) and in the Coastal Barrier Resources Act in which Congress called for preservation of the habitat value of remaining undeveloped barriers.

The New Jersey coast has been considerably altered from its natural state. Buildings line much of the shoreline; some resorts like Ocean City and Atlantic City are true urban environments (fig. 6.1). Beach erosion and storm flooding along the entire state coastline have prompted demands for shore protection for developed areas. This has resulted in the construction of numerous groins, bulkheads, and seawalls.

Many New Jersey barrier island communities date back a century or more; groins and bulkheads were constructed as early as the mid-nineteenth century. Many early protection structures were destroyed or have decayed. At present, there are over 300 groins on New Jersey's 127 mi. (205 km.) Atlantic shoreline, and about 12 mi. (20 km.) of shoreline is protected by massive seawalls. Bulkheads extend along most of the length of many developed communities. More than half of the 12 ocean inlets are controlled. Five inlets have jetties and two of the unjettied inlets are maintained by dredging. Generally, the areas with larger populations have the greater number of structures. For example, Absecon Island, the site of Atlantic City, has 40 groins (Psuty, 1984).

Effects of Human-Induced Alterations on Shoreline Change

Major alterations to coastal barriers associated with development include: (1) modification of landforms by earth-moving machinery to provide building sites; (2) construction of buildings and support infrastructure (roads, utilities); (3) emplacement of shore protection structures to protect these facilities; and (4) dredge and fill operations to augment natural beaches. These alterations are interrelated in the sense that both the desired and undesired effects of each may give rise to human adjustments which alter the barrier system further from natural conditions (table 6.1).

SHORELINE CHANGES

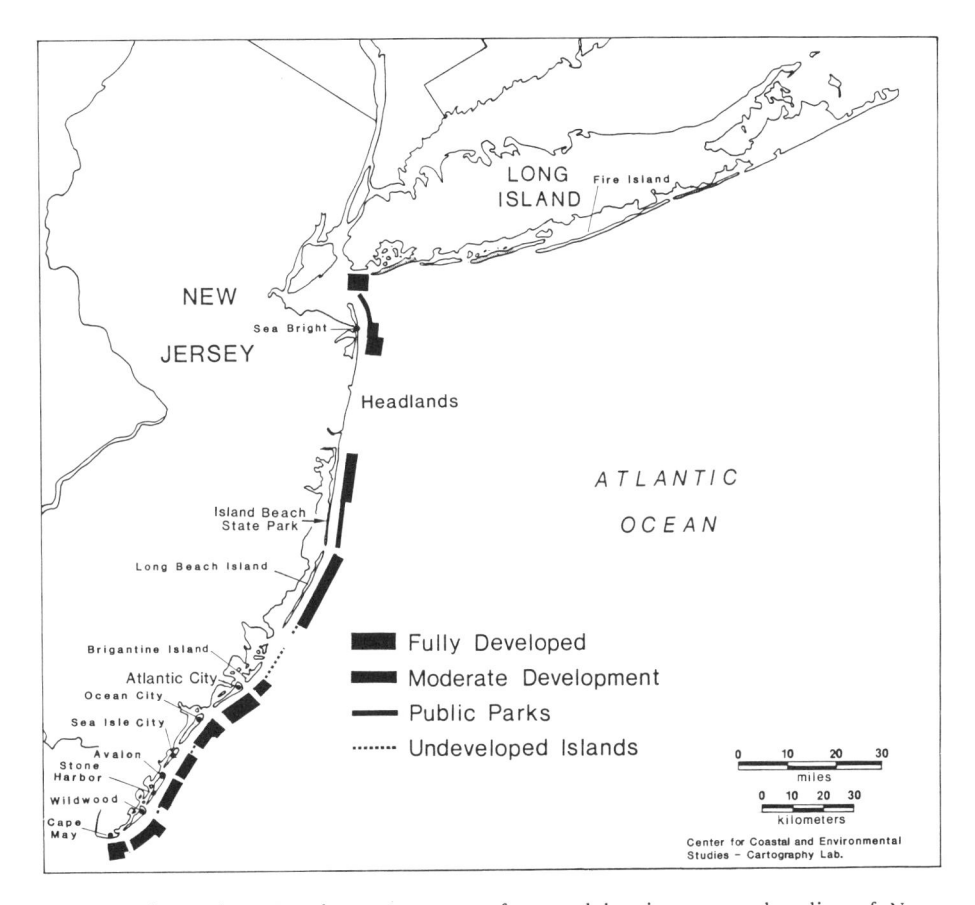

Figure 6.1. Development status of coastal barrier ocean shoreline of New Jersey

TABLE 6.1

HUMAN ALTERATIONS TO COASTAL BARRIERS WHICH AFFECT
CONFIGURATION OF LANDFORMS AND ALTER SEDIMENT BUDGETS

Human Adjustment	Direct Effects	Indirect Effects
Building site modification (grading, paving)	Changes landform configuration. Eliminates sources of sediment. Changes rates of sediment transport across land surface.	Creates disequilibrium landform configuration which can only be maintained through artificial means.
Construction of buildings	Alters wind patterns. Obstruction to sediment flow. Truncates beach or dune zone.	Focuses human use and human impacts. Leads to construction of support infrastructure and implementation of protection structures.
Emplacement of shore protection structures	Armors shoreline. Alters sediment flows. Changes location of erosion/deposition and its severity.	Leads to need for more structures.
Beach nourishment	Changes sedimentation rates and severity of erosion.	Masks nature of erosion problem. Leads to further development.
Channel dredging	Alters natural channel pattern.	Changes location of onshore erosion/deposition patterns. Leads to calls for structural improvements.

SHORELINE CHANGES

Effects of Buildings

Buildings can have considerable effects on sand transfer by wind. Condominiums in Sea Isle City, for example, create a local eddy that reverses the direction of strong westerly winds. The result is onshore transport of sand from the beach, even when the regional wind blows offshore. The sand that is deposited on the walkways in front of the condominiums must then be removed by earth-moving equipment.

Even small buildings can have pronounced effects on winds and sand transfers. A simulation study conducted on data from Fire Island, New York revealed that houses constructed on pilings occupying an average frontage (43% of the dune crest length) could reduce the natural rate of windborne sand transport by 23.7%. The same houses built on ground level with a single sand fence placed in front of them could reduce the natural rate of aeolian transport by 67.5% (Nordstrom and McCluskey, 1985).

Buildings on the beach affect sand transfers by waves. Buildings elevated on pilings cause little interference. Although buildings located on ground level obstruct sand flow, it is not known how significant this interruption is to overall barrier sedimentation. Examination of aerial photographs taken after the March 1962 storm indicated that small buildings were rapidly removed or destroyed by storm waves without causing pronounced changes in sedimentation patterns. Post-storm clean-up operations and artificial reconstruction of beaches and dunes have a far greater effect on barrier characteristics.

Most of the developed New Jersey shoreline is protected by engineering structures, and these structures have more effect on beach change than the buildings they protect. Geomorphologists have been most critical of the effects of static protection structures, so much of this discussion will be devoted to their effects.

Effects of Shore Protection Structures

A look at the New Jersey shoreline today reveals large numbers of shore protection structures with narrow beaches fronting them (fig. 6.2). It is tempting to assume that these structures have accelerated beach erosion, and several geomorphologists have made this claim. Historical shoreline change data, however, often show less erosion under conditions of development (after 1900) than do the data gathered before development (Nordstrom et al., 1977, pp. 110-112). This observation is compatible with other assessments of erosion rates on the New Jersey shoreline, which identify several barrier islands with an accretional trend (Everts and Czerniak, 1977; Dolan, Hayden, and Heywood, 1978; Galvin, 1983). The lowest rates of recent erosion are usually in locations that are protected by engineering structures. Shoreline advances in protected areas may occur at the expense of adjacent undeveloped areas, as at Stone Harbor (fig. 6.3). Accelerated erosion in undeveloped areas near engineered shorelines has been cited by geomorphologists as a reason why hard structures are not a viable option for

SHORELINE CHANGES

Figure 6.2. The New Jersey shoreline near Sea Bright/Monmouth Beach

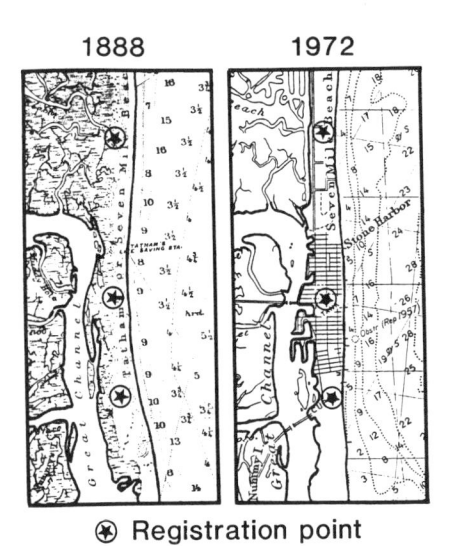

⊛ **Registration point**

Figure 6.3. Shoreline change at Stone Harbor, New Jersey

shore protection. However, some undeveloped areas next to protected shoreline segments (e.g., Northern Brigantine Island, Island Beach State Park) have not eroded at appreciably higher rates since the protection structures were built. Historic shoreline records indicate that perceptible shoreline offsets characterized these areas prior to development. We cannot identify adverse effects of protection structures simply because there is a difference in shoreline orientation in the unprotected segment.

Deleterious effects of protection structures must be distinguished from the effects of ambient natural processes and artificial beach nourishment. Accelerated erosion has occurred downdrift of some large-scale engineering structures, as at Cape May Inlet, but not at others, such as at Absecon Inlet. The shoreline fronting the massive seawall and groins in northern New Jersey at Sea Bright has been eroding rapidly, but other locations with extensive vertical structures (seawalls, bulkheads) and groin fields are not eroding appreciably.

Both Sea Bright and Cape May are at the downdrift ends of the longshore sand transport system and are in locations which have a higher net rate of longshore transport than the shoreline segments updrift of them (Caldwell, 1967). This condition has created a natural sediment deficit. Erosion at Cape May has been intensified by the jetties (designed to improve navigation) because the jetties were built updrift of the city and they increase the sediment deficit. The erosion rate at Sea Bright was reduced following emplacement of engineering works there because these were built within the segment, specifically for shore protection. Through time, the beach at Sea Bright is being eliminated, but the rate of landward displacement of the beach is lower than it was prior to construction of the seawall and groins.

Accelerated erosion on downdrift inlet shorelines, such as Wildwood, Ocean City, and Avalon is due to inlet processes rather than to shore protection structures in the vicinity (fig. 6.4). Through time, there is a natural mechanism for return of sediment to critically eroding areas. The protection structures in these locations simply buy time until natural accretion is favored once again. Thus static protection structures work well where they are a backup for the natural protection provided by a beach. Many of the bulkheads constructed following the March 1962 storm have this characteristic. They are buried beneath the dune much of the time, but they protect the buildings during storms, when the dunes are eroded.

In some locations, static structures have become a substitute for the protection provided by the beach. In areas like Sea Bright, they may be the most feasible response to shoreline erosion. Loss of the seawall there would create a breach in the barrier that probably would not close. The alternative of beach nourishment would be economically unfeasible because new sand would be rapidly removed by wave action.

There are locations where shore protection structures have increased erosion rates on adjacent shorelines, such as portions of the shoreline downdrift of the Cape May Inlet jetties. At Stone Harbor Point and the southern end of Long Beach Island, downdrift of extensive groin fields, accelerated erosion has resulted from sand starvation. The effect of this erosion is not simply the landward displacement of the shoreline with all of its characteristic features intact, as would occur under natural conditions. At Stone Harbor Point, dune building and marsh

Figure 6.4. The shoreline of Avalon, New Jersey, downdrift of Townsend Inlet, showing changes following relocation of the ebb tide channel as a result of dredging

SHORELINE CHANGES 73

Figure 6.5. Southern Long Beach Island, showing erosion downdrift of groins

construction cannot keep pace with the effects of wave attack and overwash; the result is the destruction of these features.

Rapid erosion at Stone Harbor Point and Long Beach Island has been favored by their locations adjacent to tidal inlets. Barriers like these, updrift of inlets, are characteristically low and narrow because their past growth was primarily alongshore by spit deposition rather than upward growth by dune building. Local tidal currents increase sediment mobility, which also contributes to barrier instability. Other investigators have found that erosion problems tend to be most common adjacent to tidal inlets (Kana, 1983). When human-induced sediment starvation is added to the natural vulnerability of these areas, the result is destruction of the barrier beach environment.

Problems of accentuated sediment starvation at developed inlets have profound implications for the successful achievement of the goals of the Coastal Barrier Resources Act (CBRA). Inspection of aerial photographs reveals that approximately one-third of the Atlantic Coast units designated by the CBRA are adjacent to shoreline segments which contain protection structures. Over 50% of the units are adjacent to segments that are developed with houses and where future protection structures are possible. About 70% of the units are adjacent to inlets, and nearly 30% of the total are adjacent to both inlets and protected segments. The potential thus exists for destruction of the resource potential of many of the units designated by the act.

Nourishment and Artificial Fill

In addition to static structures, there are other human adjustments which have affected sediment budgets and erosion rates. Small-scale beach nourishment operations have been conducted all along the New Jersey shoreline, and sizeable quantities of sand have been added to the beach at Ocean City, Atlantic City, and Sandy Hook. An artificial dune was constructed along a large portion of the New Jersey shoreline following the damaging March 1962 storm. Some of the larger operations have been studied, but the effects of most of the projects have not been examined.

Artificial nourishment has been used to fill breaches in coastal barriers resulting from storm wave overwash before the breaches could develop into inlets. The repair of these breaches and the replenishment of beaches in eroding areas result in less overall shoreline change than might be expected in the absence of these adjustments. It is likely that artificial nourishment has also lessened the adverse downdrift impacts to the beach caused by stabilization of the shoreline with static structures, but it is difficult to quantify these effects, given the modest amounts of fill and the lack of detailed baseline data.

Effects of Dredging

Dredging operations can have pronounced effects on shorelines far from the location of direct human alteration, because of changes to circulation patterns caused by the deepening of channels. In 1978, the main ebb tide channel at Townsend Inlet was located close to the shoreline in the northern part of Avalon (fig. 6.4). That year, a new channel was dredged through the ebb tidal delta north of the former channel to improve navigation. This alteration caused dramatic changes in the location of erosion and deposition. The new channel provided a shorter connection between the open sea and the bay and it became dominant. The current through the old channel weakened and that channel filled as a result of natural sedimentation. The beach in the inlet throat, which had formerly been eroding due to currents in the channel, became depositional, and the formerly accreting beach fronting the open ocean eroded rapidly. This sequence of events in fact mimicked natural conditions. New tidal channels are frequently created updrift of old channels during storms (Fitzgerald, Hubbard, and Nummedal, 1978). Although not designed as such, the Townsend Inlet operation indicated the possibility of establishing new patterns of sediment exchange through artful dredging. Other studies have pointed to channel dredging as a means of changing the location of accretion and erosion near inlets (Kana, 1983). These indirect effects of dredging are, as yet, poorly understood in contrast to the direct effects of dredge and fill projects.

Changes Due to Accelerated Sea Level Rise

Identification of future changes to the New Jersey shoreline cannot be projected solely from past changes. There is increasing evidence that the buildup of carbon dioxide will raise global temperatures substantially (the "greenhouse effect") and that this increase is likely to be accompanied by an accelerated rise in average sea level (Seidel and Keyes, 1983). This change will, in turn, contribute to more rapid shoreline modification than has occurred in the past. The geomorphic response of coastal barriers would be the landward and upward translation of the zone of active shoreline processes. The rate of shoreline retreat would probably be accelerated in many locations where coastal formations are not normally exposed to attack by high energy waves.

It is likely that physical changes induced by sea level rise will occur more rapidly than in the past, but they will still be gradual as viewed from a planning perspective. Projections of future geomorphic alterations due to accelerated sea level rise show considerable change in unprotected areas, but virtually no change in areas protected by seawalls (Kana et al., 1983; Leatherman, 1983). Engineering structures are therefore likely to be the preferred strategy in areas that are already developed. Even where most buildings are elevated on pilings and thus protected from increased flooding, the need to keep older, low-lying houses and roads free from flooding would likely result in the construction of vertical protection structures. Bay erosion would increase and bulkheads could be built along coastal bays to serve the dual purpose of flood protection and erosion protection. The result on many coastal barriers would be a ring levee encircling all development.

Problems caused by sediment starvation on unprotected shoreline segments will be increased as a result of sea level rise. The unprotected ends of some of the protected barrier shorelines (fig. 6.1) will be eliminated, much as Stone Harbor Point is being lost.

Non-structural alternatives are likely to be employed in undeveloped areas or developed areas where most of the buildings are destroyed by a coastal storm. Accelerated sea level rise will cause these barriers to migrate landward, and the location and configuration of beaches and dunes will change more rapidly than at present. These environments would continue to exist, but they would be more dynamic, and prudence would be required in their use.

Factors Complicating Geomorphic Analyses of Developed Areas

Isolation of specific effects of buildings or protection projects is difficult, because many adjustments have been made without prior study or follow-up investigations. In many cases, records of shore protection programs are nonexistent, sketchy, or buried in unpublished memoranda. The archival research required to identify the dates of emplacement of all protection structures and all small-scale nourishment operations in New Jersey would be a major undertaking. It would be impossible to identify all of the sand-fencing and beach-grading operations.

The specific effects of individual projects are masked because of the lag time between a human action and its geomorphic result. In the absence of subsequent human alterations, gross changes would be possible to determine because the response of unconsolidated sediments to wave action is rapid. But continuous human adjustments (new buildings, groins, dredging, beach grading, nourishment) constantly change physical conditions; geomorphic investigations of the effects of each project suffer from a lack of baseline control.

Identification of the effects of human adjustments on natural systems is a challenging task. The magnitude of the problem of isolating cause and effect will deter some geomorphologists from researching developed coastal barriers. Many geomorphologists simply do not want to study human-altered systems. Those involved in modern process research seek undisturbed areas as field sites and often add human-induced effects as an overlay to the natural system. Many systems have only recently been disturbed, so that there is an insufficient data base with which to document change. Assessment of the future impacts of today's changes on the environment will require the study of systems that have already been heavily disturbed. These disturbed systems are the best research base for the identification of future land-use scenarios (Nordstrom, 1984).

The effects of development on coastal barriers are easier to conceptualize than to substantiate. In New Jersey, it is difficult to provide conclusive geomorphic evidence to suggest that development (buildings and shore protection structures) will categorically increase erosion rates, at least within the time scale considered by planners, engineers, and developers. Some areas will be more susceptible to the adverse effects of development than others, but mitigation of these effects is both possible and likely. It is difficult to identify all of the natural and human causes of change, let alone to predict the future changes.

Conclusion: Future Research Needs

Statements that residential construction is incompatible with coastal barrier processes in the long term (hundreds of years) are not taken seriously by prospective buyers and developers who seek a short-term return on their investment. Geomorphologists should focus more attention on the mid-range (10-50 year) effects of development. The existence of buildings on a coastal barrier need not interfere significantly with the integrity of the natural features or with the natural transfer of sediment along and across the barrier. Houses may be elevated above areas where sand is moved by wind and waves, for example. Dunes need not be leveled or paved over for house construction. Occupation of a shorefront home does not require paved roads, extensive landscaping, or other needs of a home in "suburbia." Criteria for the construction and use of buildings can be designed to be more compatible with a barrier beach location. Potential threats to the integrity of the remaining undeveloped coastal barriers caused by poorly designed development underscores the need to design and implement these criteria.

There are still too many unknowns to be able to make categorical decisions to exclude development on coastal barriers that are not presently developed. Few would argue, however, that no control of development is a responsible policy. Future research should seek to determine the degree of development that can occur on coastal barriers without causing severe adverse impacts. The results of that research can then be integrated with the results of studies designed to minimize losses of life and property by storm damage. The integration of these two areas of geomorphic research can then provide a basis for future decision on the management of developed coastal barriers.

Acknowledgments

This report is partially based on research sponsored by the National Sea Grant College Program, NOAA, U.S. Department of Commerce Project No. RS-3, Grant No. NA81AA-D-00065. I would like to thank Paul A. Gares, Norbert P. Psuty, and Lester B. Smith, Jr. for helpful discussions about the effects of development on barrier beach processes.

References

Adams, J.W.R. 1981. "Florida's Beach Program at the Cross-Roads." *Shore and Beach* 49(2): 10-14.

Caldwell, J.M. 1967. "Coastal Processes and Beach Erosion." Fort Belvoir, VA: U.S. Army Corps of Engineers, Coastal Engineers Center Reprint 1-67.

Dolan, R., B. Hayden, and J. Heywood. 1978. "Analysis of Coastal Erosion and Storm Surge Hazards." *Coastal Engineering* 2: 41-54.

Everts, C.H., and M.T. Czerniak. 1977. "Spatial and Temporal Changes in New Jersey Beaches." In *Coastal Sediments '77.* New York: American Society of Civil Engineers, pp. 444-459.

FitzGerald, D.M., D.K. Hubbard, and D. Nummedal. 1978. Shoreline Changes Associated with Tidal Inlets along the South Carolina Coast." *Coastal Zone '78.* New York: American Society of Civil Engineers, pp. 2684-2705.

Galvin, C.J. 1983. "Sea Level Rise and Shoreline Recession." In *Coastal Zone '83. Ed. O.T. Magoon and H. Converse.* New York: American Society of Civil Engineers, pp. 2684-2705.

Kana, T.W. 1983. "Soft Engineering Alternatives for Shore Protection." *Coastal Zone '83.* New York: American Society of Civil Engineers, pp. 912-929.

Kana, T.W., J. Michel, M.O. Hayes, and J.R. Jensen. 1983. "Shoreline Changes Due to Various Sea-Level Rise Scenarios." *Coastal Zone '83.* New York: American Society of Civil Engineers, pp. 2890-2901.

Leatherman, S.P. 1983. "Geomorphic Effects of Projected Sea Level Rise: A Case Study of Galveston Bay, Texas." In *Coastal Zone '83. Ed. O.T. Magoon and H. Converse.* New York: American Society of Civil Engineers, pp. 2890-2901.

Nordstrom, K.F. 1984. "Assessing Future Changes." *Geotimes* 29(1): 4-5.

Nordstrom, K.F., et al. 1977. *Coastal Geomorphology of New Jersey.* New Brunswick, NJ: Rutgers University Center for Coastal and Environmental Studies Technical Report, NJ/RU-OCZM-TR No. 1-300.

Nordstrom, K.F., and J.M. McCluskey. 1985. "The Effects of Houses and Sand Fences on the Eolian Sediment Budget at Fire Island, New York." *Journal of Coastal Research* 1: 39-46.

O'Brien, M.P., and J.W. Johnson. 1980. "Structures and Sandy Beaches." In *Coastal Zone '80. Ed. B.L. Edge.* New York: American Society of Civil Engineers, pp. 2718-2740.

Parker, N. 1980. "Structural Approach to Coastal Problems." In *Barrier Island Forum and Workshop.* Ed. B.S. Mayo and L.B. Smith, Jr. Boston: National Park Service, North Atlantic Region, pp. 127-138.

Pilkey, O.H. 1981. "Geologists, Engineers, and a Rising Sea Level." *Northeastern Geology* 3: 150-158.

Psuty, N.P. 1984. "Shoreline Structures of New Jersey and New York." In

Artificial Structures and the Shoreline. Ed. H.J. Walker. International Geographical Union Commission on Coastal Environments Publication (in press).

Seidel, S., and D. Keyes. 1983. "Can We Delay a Greenhouse Warming?" United States Environmental Protection Agency Report EPA-230-10-84-001. Washington: EPA Strategic Studies Staff Office of Policy Analysis, Office of Policy, Planning and Evaluation.

Vaccaro, M.J. 1981. "New Jersey Seashore - Ultimate Destruction or Salvation." *Shore and Beach* 49: 34-37.

PLANNING AND GROWTH MANAGEMENT

MANAGING CHANGE ON DEVELOPED COASTAL BARRIERS

David J. Brower and Timothy Beatley

University of North Carolina at Chapel Hill

Introduction

Barrier beaches, like other coastal areas, are attractive places in which to reside or vacation. Increases in personal income and recreation time have fueled demand for coastal barrier development. Changes in such communities have often been drastic, unexpected, and unplanned. Most barrier communities are small and their capacity to manage change is usually low. The town of Long Beach on Oak Island in North Carolina, for example, has increased in population by over 1700% since 1960 (Town of Long Beach, 1984).

The development strain placed upon barrier resources is considerably greater than would be indicated from permanent population figures. For example, while the town of Emerald Isle, North Carolina has a permanent population of about 865, its average seasonal population is estimated at around 8,500 (Town of Emerald Isle, 1984).

Even though substantial growth and development has occurred on many coastal barriers, the political and social systems in place are still often "small-town" in character. (Long-established cities like Galveston and Miami Beach of course are exceptions.) County governments are often responsible for managing growth in unincorporated coastal barrier communities. The local economies of barrier communities are dependent on tourism and recreation-oriented industries, and coastal governments are typically anxious to ensure the health of these sources of revenue.

Different political circumstances are likely to yield varying levels of support or opposition for growth management, as well as the different objectives behind programs that are enacted. The political dynamics of coastal barrier island communities require growth management policies that are sensitive to them. By growth management we refer to public programs and policies which control or influence the type, quantity, location, density, quality, or timing of development that occurs in a jurisdiction. The following section examines some of the major competing reasons why growth management programs are, or could be, adopted.

Goals of Coastal Barrier Management

Many of the goals of growth management programs are implicit rather than explicit and need to be brought to the surface for more adequate debate. A clear set of objectives, even if left at a relatively general level, is essential for determining the type and content of growth management programs most suitable for developed coastal barrier communities. The following sections present several of the most important types of these goals.

Environmental Quality

Coastal barriers represent extremely sensitive and complex ecological systems. Growth management programs can be designed to reduce the adverse impacts of new development upon these systems. A central component of the North Carolina Coastal Area Management Program, for example, is the designation of "Areas of Environmental Concern" (e.g., coastal wetlands, estuarine shoreline, public trust waters) and the regulation of new development to lessen its impact on the environment.

Protection of Scenic and Aesthetic Qualities

Much of the aesthetic appeal of coastal barriers is a function of human interaction with this natural environment. For instance, building height restrictions are sought in some communities to lessen their perceptual impacts, such as the loss of views toward the ocean, the loss of afternoon sunshine, and the perceived ugliness of such structures. Development of coastal barriers entails such questions as the configuration and interaction of different types of human uses, such as transportation networks, public and private spaces, and the overall sense of community and "livability" generated by these land-use patterns (Lynch, 1981; Lynch and Hack, 1983).

Efficiency and Fiscal Objectives

Growth management programs may aim to ensure that public funds are spent in responsible and efficient ways. Local governments may control the expenditures for public services so as to maximize benefits and minimize costs. The location and timing of the construction of new roads, sewers, and water lines are prime examples. Substantial public savings can be realized if the provision of these facilities is coordinated with private development decisions and vice versa. Subdivision exactions, user fees, or impact taxes on new development may supplement property taxes as revenue sources to provide such facilities (Stroud, 1978; Mercer and Morgan, 1981).

Employment and Economic Vitality

Growth management may involve efforts to enhance economic conditions so as to increase employment and income levels in the community. Many, if not most, coastal barrier economies are highly seasonal and create substantial employment disruptions for year-round residents. Also, many barrier economies lack the diversity that ensures stability. A community heavily dependent on tourism may be economically devastated should the price of gasoline rise dramatically or the amount of disposable personal income shrink. Growth management policies may enhance the local economy by encouraging new types of commerce and industry.

Equity and Accessibility

Growth management programs may also seek to ensure the accessibility of barriers to all social classes and income groups. Accessibility may be advanced by ensuring that adequate housing exists for all individuals employed in these communities, and through adequate public parking for visitors. Zoning and land use regulations are typically employed to reduce what are perceived to be the unfair, negative impacts of the actions of some on the welfare of others and the community in general (Beatley, 1984). Mobile homes may be required to be tied down, for instance, because they can act as battering rams, destroying other homes in the community should a hurricane strike. Furthermore, growth management programs may assess individuals for the inevitable costs and demands they impose on the community.

Public Health and Safety

Growth management programs have traditionally sought to protect public health and safety, e.g., efforts to reduce noise, traffic and congestion, the isolation of noxious or dangerous activities, etc. On coastal barriers this goal is perhaps most evident with respect to planning for hurricanes. Growth management policies such as at Sanibel, Florida can restrict development according to the capacity of hurricane evacuation routes. Provision for adequate storm shelters or "vertical evacuation" in new high-rise developments are also important goals.

Carrying Capacity as an Organizing Concept

The concept of "carrying capacity" is useful in the formulation of a growth management program. The basic idea is that there are thresholds--environmental, social institutional--of growth and development, beyond which certain unacceptably large negative effects will result, unless public actions are undertaken to increase the carrying capacity (Godschalk, Parker, and Knoche, 1974; Schneider, Godschalk, and Axler, 1978). At an ecological level, we can establish environmental thresholds for air and water quality, wetlands protection, sanitary sewage disposal, and so on. These thresholds may be translated into density of development or number of dwelling units which the barrier, or portions of it, can sustain with existing services. Carrying capacity might also be applied to aesthetic issues, such as the scenic qualities of the coastline, and the feelings of appropriate scale and intimacy. A threshold regulating the overall mass, height, and configuration of structures in a town's central business district might be established. Other thresholds may be formulated for traffic flow, public usage of beaches and nature areas, and the amount of low- and moderately-priced housing in the community.

Another important application of carrying capacity on coastal barriers is the ability of local road and bridge systems to accommodate timely evacuation in the event of a hurricane (Brower, Einsweiler, and Propst, 1984; Clark, 1976). Carrying capacity offers a promising approach to implementing general community goals and for logically structuring the growth management decisions that must be made to reach these goals.

Alternative Management Techniques

There are numerous growth management techniques that can be used to implement the carrying capacity thresholds that might be established. We may briefly identify five categories of such techniques which can be put to use in developed coastal barrier communities: (1) development regulation, (2) land acquisition, (3) capital facilities policies, (4) taxation, financial, and other public incentives, and (5) information distribution and dissemination (Brower et al., 1984). Table 7.1 provides a listing of some of the more common management tools in each of these categories.

While many coastal communities tend to rely heavily on one or two techniques, such as zoning and subdivison registration, ideally these devices should be organized into an effective package or growth management "system" which simultaneously addresses multiple issues and problems. This requires a uniquely

TABLE 7.1

GROWTH MANAGEMENT TOOLS AND TECHNIQUES

Development Regulation	Conventional zoning
	Exclusive agricultural or nonresidential zones
	Conditional and contract zoning
	Bonus and incentive zoning
	Interim or temporary development regulations
	Floating zones
	Performance zoning
	Planned unit development
	Subdivision regulations
	Cluster or average density zoning
	Environmental impact ordinance
	Annual permit limits
	Building codes
Land and Property Acquisition	Fee simple acquisition
	Less-than-fee simple acquisition
	Advance site acquisition
	Land banking
	Compensable regulation
Capital Facilities Policies	Capital improvements programming
	Urban and rural service areas
	Annexation
	Development timing
Taxation, Financial, and Other Incentives	Impact taxes
	Use-value and preferential tax assessment
	Site-value taxation
	Land gains taxation
	Public service pricing policies
Information Distribution and Dissemination	Real estate disclosure provisions
	Posting of hazard zone signs
	Construction practices seminars
	Hazard zone delineations on plats and deeds

TABLE 7.2

ARGUMENTS AGAINST GROWTH MANAGEMENT PROGRAMS

Rank	Frequency (considered at least somewhat important)	Percent
1. Growth management measures lead to increased developmental costs N=382	324	84.8
2. Decisions about risks from coastal storms are best left to the individual N=359	254	71.0
3. Growth management measures dampen local economy N=368	252	68.5
4. Particular development measures are illegal or unconstitutional N=351	232	66.1

SOURCE: Beatley, Brower, Godschalk, and Rohe, 1985.

crafted program adapted to the ecological, cultural, political, and economic characteristics of the particular barrier community.

Building Management Capacity—Overcoming Impediments

In most coastal barrier communities, application of the growth management concept is still in its infancy. In many such communities, specific techniques are either strongly resisted, considered but deemed inapproprite or infeasible, or simply not considered at all. We will briefly discuss the most significant impediments to the successful use of growth management techniques and in turn tender some thoughts on overcoming these impediments.

Private Property and Personal Freedom

Perhaps the most substantial impediment to the adoption of growth management provisions in coastal barrier communities is the mythical absolute freedom of property owners to do as they wish with their land. In a recent survey of coastal communities in 19 states, the four objections to growth management listed in table 7.2 were indicated by a majority of the respondents.

Arguments that growth management provisions increase the costs of development are the most frequently cited, followed by arguments that decisions concerning personal risk should be left to the individual. Questions about the effects of growth management measures on the local economy and the legality of these measures also appear to be important. Proponents of growth management in coastal communities must begin to acknowledge the significance of these objections.

Arguments against growth management programs are most effective when they go unanswered. They can only be defused by dealing with them explicitly and advancing compelling counter-arguments. For instance, advocates might defend the increase in development costs associated with some growth management programs (e.g., impact taxes or subdivision exactions) on the grounds of equity between old and new residents.

Limited Financial Resources

Coastal barrier communities often have limited financial resources. Funds are often scarce for expensive studies or for programs that are expensive to implement, such as land acquisition. Even if a regulatory program should be enacted, the community may not have the funds to ensure compliance.

A community can overcome limited financial resources by coordinating its programs with adjoining jurisdictions and by harnessing available outside resources such as federal disaster assistance funds in the aftermath of a hurricane. The effectiveness of local efforts may largely depend upon the extent of outside assistance.

Limited Expertise

A major obstacle to growth management in many coastal communities is the lack of professional expertise in land-use planning and management. This impediment is closely tied to a scarcity of financial resources, as discussed above. Proponents of growth management should stress the high community returns from relatively inexpensive investments in planning staff. One innovative approach is the use of "circuit rider" planners, in which a number of jurisdictions may share the cost of a trained professional. State technical assistance and training programs for local personnel are also useful.

Inexperience and Uncertainty

Resistance to new approaches may be simply due to unfamiliarity with growth management programs. Overcoming tradition and stirring a sense of innovation are important aspects of building local management capability.

Managing Reconstruction Following a Hurricane

Reconstruction following hurricanes and severe coastal storms provides an opportunity to apply growth management policies to mitigate the impact of future hurricanes as well as to achieve other goals (McElyea, Brower, and Godschalk, 1982). In this situation, growth management shifts from its traditional focus on new construction to that of rebuilding areas that have been changed or destroyed. Severe storm damage may present a coastal community with options regarding whether to rebuild a road or public structure in the same location or in the same way. The community must decide whether to discourage rebuilding in hazardous areas, and if so, how? In short, communities stricken by a hurricane may opt to change the location, quality, and nature of redevelopment—and hence the "face" of the community. In discussing the concepts that follow, we will draw heavily upon some recent experiences in planning for hurricanes for the town of Nags Head, North Carolina (Brower et al., 1984).

Coastal localities in North Carolina are required to develop reconstruction plans under the state's Coastal Area Management Program. These reconstruction components of local land-use plans may prescribe procedures for assessing damage after the storm, establish a damage assessment team, and propose the appointment of a "recovery task force" to oversee reconstruction. Communities must establish standards for guiding repairs and reconstruction. The North Carolina provisions are notable for their requirement that coastal localities plan in advance of these reconstruction opportunities, thereby increasing their capacity to manage change in what is usually a turbulent period of time.

We now describe four strategies that can be employed to organize reconstruction after a storm so that decisions about redevelopment will be more manageable.

Post-Storm Assessment

An initial requirement is to assess the actual damages caused by the storm on the natural and built environments. Where are the most severely damaged areas? What were the causes of such damage? How did natural systems respond to storm forces, e.g., did any new inlets breach the barrier island? In what condition are public facilities and infrastructure such as roads, and sewer and water lines? A group of qualified officials should be prepared to conduct the assessment. Such a team should ideally include the municipal or county planning director, public works director, building inspector, tax assessor, town engineer, and so on.

Recovery Task Force

The establishment of a recovery task force is one approach to making sense of the special decisions that must be made during this period. Such a body, likely a combination of citizens and public officials, would be given the charge of evaluating damage assessment reports, reviewing pre-storm plans and objectives, and proposing post-storm actions and policies that will capitalize upon the opportunity to mitigate future storm damage. For example, the Nags Head post-storm reconstruction plan specifies that rebuilding of public and private structures in high-hazard zones should be discouraged, that any reconstruction should conform to the State Building Code, that structures should be required to be set back a certain distance from the ocean until the Coastal Area Management Act setback line is reestablished, and so forth. While these provisions are not site-specific, they would at least provide guidance during reconstruction of the damaged areas.

While the task force is designed to oversee and manage redevelopment, its function must also include preparing to deal in advance with a possible storm event. While the formulation in advance of very detailed plans is neither feasible nor desirable (because the nature and extent of possible damage is unknown), reconstruction policies may be formulated at a general level. Preparing such contingency plans in advance will probably increase public acceptance of post-storm actions by town officials.

Delineation of Damage Zones: The "Triage" Concept

Following a severe storm, public officials must decide where to direct attention and funds. One approach is to organize the disaster problem according to the severity of damage by zone within the community. A three-tiered delineation, as suggested by Haas, Kates, and Bowden (1977), would be analogous to the concept of the "triage" used in emergency medicine. It suggests that there will be areas of "minor" damage where rebuilding should be permitted without further public consideration. In "major" damage areas, however, the community should suspend redevelopment while it considers appropriate actions. In the middle are areas of "moderate" damage where, at least initially, the bulk of public attention should be directed. The Nags Head Post-disaster Reconstruction Plan identifies and defines three such zones.

Temporary Reconstruction Moratoria

The basic purpose of a temporary reconstruction moratorium is to afford time for a locality to consider more carefully the options available to it. Should it permit redevelopment in high-hazard areas that are likely to be destroyed again in the next storm? Should density limits or land-use designations be changed before redevelopment is permitted? Should the public purchase land and damaged properties to advance community objectives such as enhancing beach access and providing open space and/or recreational areas? A moratorium can provide sufficient time for adapting reconstruction plans prepared in advance of the disaster to the specific circumstances at hand. Where such plans do not exist, a moratorium also provides an opportunity to review the redevelopment options and opportunities that such a plan implies.

While the political and legal viability of a redevelopment moratorium depends on it being temporary, it can be extended as circumstances merit. For instance, an initial moratorium might be extended if certain information crucial to decisions about redevelopment (e.g., determining the size and location of hurricane overwash zones) is not yet available. Moreover, as the "triage" concept suggests, the imposition of moratoria may be geographically variable. A moratorium in one part of the community might be lifted, while one imposed in another sector might be extended, depending upon circumstances.

Conclusions

Coastal barrier communities will be subjected to continued pressures to develop and redevelop in the years ahead and must be prepared to manage competently these pressures. Growth management measures should be used by these communities in wrestling with the inevitable problems created by extensive coastal barrier development. The types of growth management techniques used, and the ways in which they are used, will fundamentally depend on the goals that these communities wish to pursue. Some of these goals will be conflicting and will require public debate and careful prioritizing. Carrying capacity is a useful concept for understanding the nature and extent of problems generated by development, and it provides a way of concretely implementing the objectives deemed important by the locality. Growth management programs can be used effectively in guiding redevelopment following a hurricane or severe coastal storm. Increasingly, storm events will represent significant opportunities for advancing community goals and coastal barrier communities must cultivate the ability to manage effectively this redevelopment process.

References

Beatley, T. 1984. "Applying Moral Principles to Growth Management." *Journal of the American Planning Association* 50(4):459-469.
Beatley, T., D.J. Brower, D.R. Godschalk, and W.M. Rohe. 1985. *Storm Hazard Reduction through Development Management: Results of a Survey of Gulf and Atlantic Coast Communities.* Chapel Hill: University of North Carolina Center for Urban and Regional Studies, preliminary report, January.
Brower, D.J., et al. 1984. *Developing a Growth Management System.* Chicago: Planner's Press.
Brower, D.J., L.D. Einsweiler, and L. Propst. 1984. *Carrying Capacity Analysis, Town of Nags Head.* Chapel Hill: Coastal Resources Collaborative, Ltd.
Clark, J. 1976. *The Sanibel Report: Formulation of a Comprehensive Plan Based on Natural Systems.* Washington: Conservation Foundation.
Godschalk, D.R., F.H. Parker, and T.R. Knoche. 1974. *Carrying Capacity: A Basis for Coastal Planning.* Chapel Hill: University of North Carolina Center for Urban and Regional Studies, June.
Haas, E., R. Kates, and M. Bowden (eds.). 1977. *Reconstruction Following Disaster.* Cambridge: MIT Press.
Lynch, K. 1981. *Good City Form.* Cambridge: MIT Press.
Lynch, K., and G. Hack. 1983. *Site Planning.* Cambridge: MIT Press.
McElyea, W., D.J. Brower, and D.R. Godschalk. 1982. *Before the Storm: Managing Development to Reduce Hurricane Disasters.* Chapel Hill: University of North Carolina Center for Urban and Regional Studies.
Mercer, L.J., and W.D. Morgan. 1981. *City and County User Changes in California: Options, Performance and Criteria.* Berkeley: University of California Institute of Government.
Schneider, D.M., D.R. Godschalk, and N. Axler. 1978. *The Carrying Capacity Concept as a Planning Tool.* Chicago: PAS Memo, APA Report 338.
Stroud, N. 1978. "Impact Taxes: The Opportunity in North Carolina," *Carolina Planning* 4(2):20-27.
Town of Emerald Isle, North Carolina. 1984. *Storm Hazard Mitigation Plan and Post-Disaster Reconstruction Plan.* Prepared by George Eichler and Associates, and Satilla Planning, Inc.
Town of Long Beach, North Carolina. 1984. *Hurricane Plan.*

COASTAL BARRIER PROTECTION AND MANAGEMENT: COLLIER COUNTY, FLORIDA

Mark A. Benedict

The Conservancy, Inc.
Naples, Florida

Introduction

Over the last decade coastal barriers have gained wide recognition for their value, both as a natural resource and a community resource. Coastal barriers serve as the continent's first defense against storm waves and oceanic flooding. As dynamic units composed of sand, vegetation, and shallow bays, they offer storm protection to interior areas. A major storm striking communities upland of undisturbed coastal barriers would inflict fewer losses there than in those communities perched on or behind highly altered barriers.

Coastal barriers contribute to ecosystem productivity. The wetlands and shallow bays sheltered by these depositional features form the basis of the coastal food chain. Saltwater wetlands, which include supratidal marshes, intertidal mangrove forests, and subtidal marine grassbeds, are some of the most productive ecosystems in the world. The organic materials produced by these plant communities constitute the first level of the estuarine and nearshore marine food chain. Without the sand beach and the upland back-barrier zone, these productive areas could not exist along most of our coasts.

Coastal barriers also contribute significantly to the biological diversity of the coastal zone. The sheltered, nutrient-rich habitats found across these land forms provide a niche for many species of plants and animals. Coastal barriers are a valuable resource for local communities. They are the site of shorefront recreation activities which include fishing and shellfishing, swimming, sunbathing, shell collecting, bird-watching, and nature study. Their wetlands power the marine food chain which is harvested commercially. Their aesthetic appeal, which draws both tourists and new residents to the area, contributes to the economies of the local communities.

The coastal barriers of the U.S. Atlantic and Gulf coasts have been the site of intense land development. Shoreline construction too close to the shore has resulted in the loss of the protective and recreational value of many miles of sandy beaches. The conversion of wetlands to developed uplands has resulted in a reduction of coastal productivity and a decrease in harvests. The destruction of natural habitats has reduced biological diversity in many sections of the coastal zone. Furthermore, the concentration of human populations on these low, coastal landforms has placed the lives and property of residents in danger, should a major coastal storm strike.

Over the past few years there has been an increased emphasis on the management of coastal barriers in order to protect, maintain, and restore resource values, as well as to provide for the safety of residents. At the federal level, the passage of the Coastal Barrier Resources Act of 1982 set the stage for reforms in coastal barrier management policies by prohibiting the use of federal funds for the development of unaltered Atlantic and Gulf Coast barriers (see Godschalk, this volume). Many states similarly are looking into ways to restrict barrier development while still permitting a reasonable use of property. Florida has developed guidelines for coastal construction and has encouraged counties and municipalities to adopt them as standards.

Federal and state legislation is, however, not enough. Local governments must also take an active role in the protection and management of coastal barrier resources, because they have the most to lose from poorly planned development. Only at the local level can community goals be set and strict, site-specific management policies be implemented for the protection of natural resources.

The Collier County Coastal Barrier Program

Collier County provides an excellent example of a local government's attempt to protect coastal barrier resources while allowing for a reasonable use of private property. Collier County shared the intense development that occurred along the Atlantic and Gulf coasts during the 1960s and 1970s. Coastal barriers in the vicinity of Naples and Marco Island have been significantly altered. Yet, of the 39 miles of coastline in Collier County, 61% remains relatively undisturbed. The majority of these undeveloped coastal areas are under county jurisdiction. The population of Collier County is increasing, and the coastal zone's proximity to the county's major recreational resources make it the number one housing choice for new residents. Because of these factors, there is increasing pressure to develop those coastal barriers presently unaltered by coastal land-use activities.

Realizing the value of barrier resources and the lack of specific criteria to evaluate proposed coastal zone activities, the county's Natural Resource Management Department is in the process of developing a resource management and protection program for its coastal barriers. The development of this program is being funded by the Board of County Commissioners, the federal Office of

COASTAL BARRIER PROTECTION AND MANAGEMENT

Coastal Zone Management (OCZM), Ocean and Coastal Resource Management (OCRM), and the Florida Department of Environmental Regulation.

The Coastal Barrier Program is the first detailed resource management program developed pursuant to the Collier County Natural Resource Management Plan (Benedict et al., 1984a). It is also the first phase of a county Coastal Zone Management Program that will ultimately cover the barrier islands, estuaries, and saltwater wetlands of Collier County's coastal zone. The Coastal Barrier Program involves a multi-faceted approach to natural resource management. It recognizes the need for management based on system-wide as well as site-specific resource data, and it covers all coastal barrier zones, their functions, and the processes that influence them. As such it is designed to provide firm standards for coastal barrier activities yet take into consideration site-specific data that could influence management decisions. The Coastal Barrier Program is currently composed of the seven components explained below.

Coastal Barrier Resource Inventory and Historical Analysis

The essential first step in any natural resource management program is the collection of ecosystem-wide data on the resources to be protected and the dynamic processes that influence them. This approach is commonly used by the National Park Service in the management of their parklands (Godfrey and Benedict, 1977). During the early months of program development, the Natural Resources Management Department undertook a resource inventory of the county's coastal barrier, which yielded county-wide information on the current status of both natural and artificial coastal barrier features. Specifically, the resource inventory provided data on geomorphology, physiograhic zonation, biota, land use, structural stabilization, and beach access.

Information on existing coastal barrier resources is not sufficient to make enlightened management decisions. An understanding of beach processes and changes in the coastal barriers over time is also necessary. A historical analysis was therefore conducted to provide information on climate and hydrography, littoral drift, sand supply, beach profile changes, shoreline changes, and tidal pass dynamics (Harvey et al., 1984).

The resource inventory together with the historical analysis provide the coastal barrier manager with a powerful tool, enabling a predictive capability. Although nothing can be forecast with certainty, this is the best and perhaps the only way for the resource manager to look into the future while reviewing applications for present day activities.

Coastal Barrier Units

The coastline of Collier County varies in geomorphology, shoreline dynamics, ownership, and land use. Although certain permit standards can and should be applied county-wide, differences in coastal barrier status necessitates some variation in management policy. Rather than trying to create general standards applicable to a wide range of conditions, it is best to tie them to recognizable management units. Nine such units have been identified in Collier County (fig. 8.1). Separated from one another by tidal passes, they represent discrete entities where coastal processes and ecological function are closely linked.

The nine coastal barrier units are further subdivided into distinct beach segments, utilizing data obtained from the aforementioned resource inventory and historical analysis. These beach segments are delineated according to their resource features and their predicted shoreline changes (fig. 8.2). The technique of identifying and separating beach segments for management purposes has been applied previously by Pilkey et al. (1978) and Harvey (1982). Site-specific information on beach/barrier characteristics, shoreline migration history, recreational values, and hazard potential emphasizes the distinct nature of different portions of the coastline and the need to treat each separately for management purposes.

All coastal barrier unit and beach segment data have been recorded on standardized data forms (Benedict et al., 1984b). The data will be entered into a computerized data base management system to enable easy access to and expansion of the information base. Tied to these coastal barrier data forms is a Coastal Management Atlas (Benedict et al., 1984c) which contains beach segment air photos detailing sampling transects, stations, beach access points, shoreline stabilization structures, and a wide array of other beach segment data.

Coastal Barrier Zones

Resource value is closely tied to ecological characteristics. Any plan for the management and protection of coastal resources must therefore be based on existing ecological zones. Five such zones are recognized for the coastal barriers of Collier County: the nearshore, active beach, dune/washover, stabilized backbarrier, and wetland zones. Each zone possesses distinct ecological features and functions, yet all are interconnected as parts of the Southwest Florida coastal barrier ecosystem.

In order to function, each zone must retain certain biological and physical features. Removal of these features destroys a zone's function and degrades the associated resource value. Therefore, an essential role of the coastal barrier program is to identify those resource values that need to be protected through management, and to describe the minimum conditions needed to maintain those values. As an example, the value of a natural sand beach in storm protection is widely recognized. Beach profiles shift in response to wave energies and so retain a dynamic balance between sand supply and sea level while dissipating the strength of wave attack. To function correctly, a minimum, obstruction-free, active beach must be maintained so that these processes can occur naturally during storm periods. When artificial shoreline structures such as seawalls or revetments have been placed in this active zone, an accelerated deepening of the shorefront profile occurs, resulting in a loss of the open recreational beach. If storm protection and recreation are identified as resource values that need to be retained, the correct management actions must be employed to maintain minimum active beach features.

COASTAL BARRIER PROTECTION AND MANAGEMENT 99

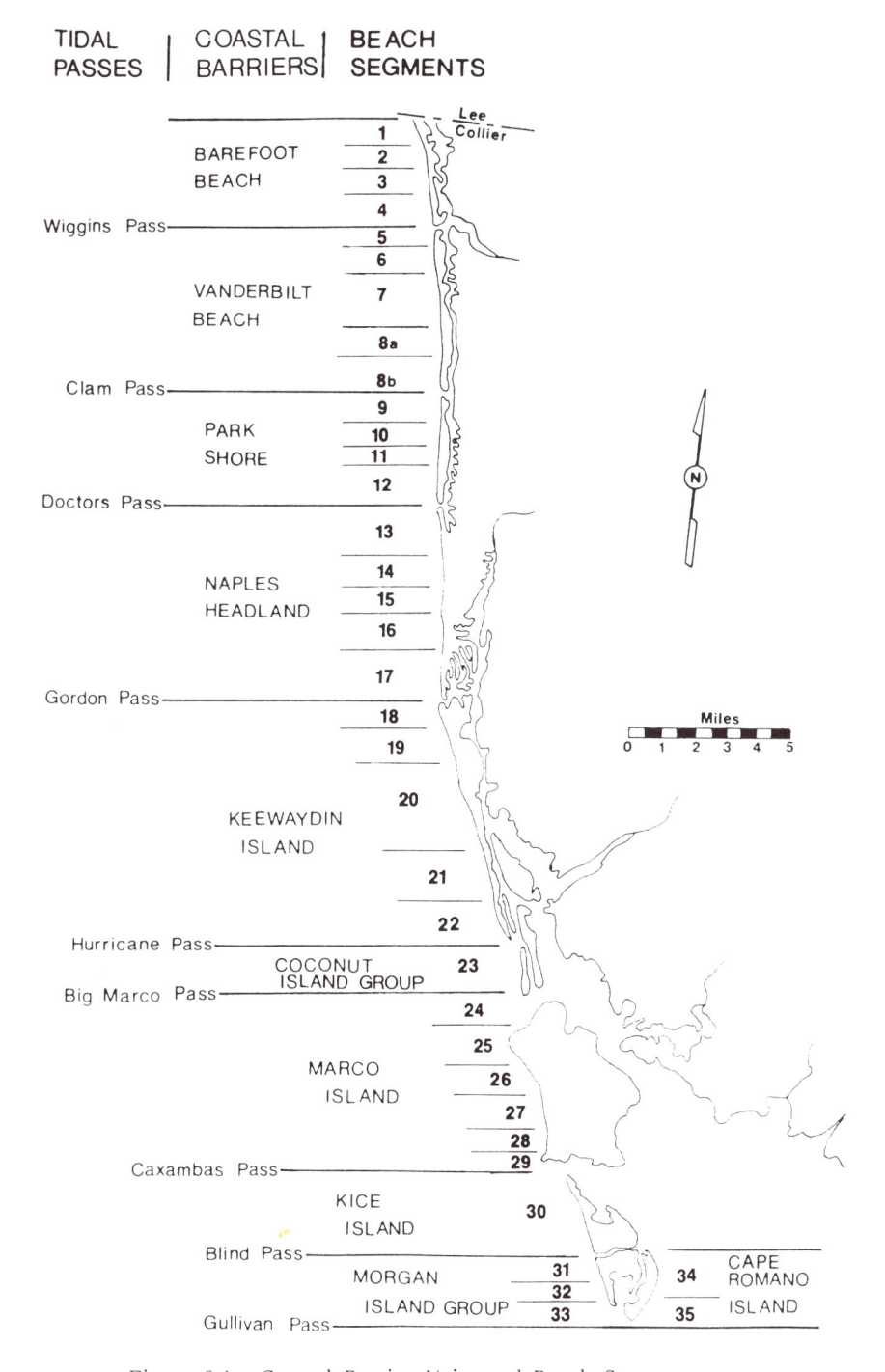

Figure 8.1. Coastal Barrier Units and Beach Segments

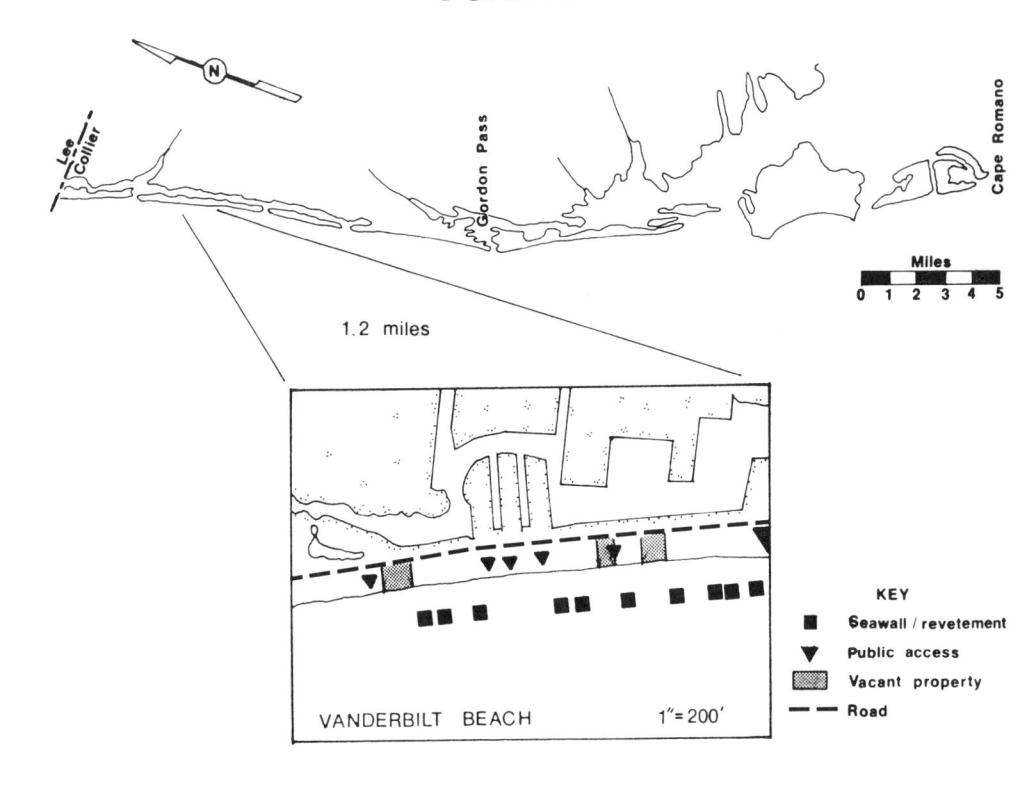

BEACH SEGMENT CONCEPT

KEY
- ■ Seawall / revetement
- ▼ Public access
- ▒ Vacant property
- — — Road

VANDERBILT BEACH 1"= 200'

- Coastal Barrier Type
- Shoreline Migration History
- Beach / Dune Characteristics
- Land Use & Ownership
- Recreational Value
- Management Considerations

Figure 8.2. Beach Segment Concept

Coastal Barrier Activities

Once resource values are identified, it then becomes necessary to classify all potential activities and to assess their impacts on the ecological characteristics of the different barrier zones. A wide variety of activities can occur on coastal barriers; for convenience these are grouped as alteration, construction, devegetation, recreation, and restoration. These activities range from those with low or beneficial impacts to those that obliterate the function of a given zone and its resource values. An understanding of the types and impacts of all potential activities is necessary for an effective coastal barrier management program.

Land-use Matrices

Land-use matrices are the most effective tool for correlating resource values with the impacts of various activities (Brown and Starnes, 1983). A matrix is prepared for each of the activity categories and then, using research data, all activities in that category are classified according to their probable impact on a specific zone and its resource values. In the land-use matrix, low impact or beneficial activities are classified as "compatible" with the features of a given zone while activities that degrade a zone are denoted as "incompatible." Although a small number of coastal barrier activities fall clearly into one or the other of these two classes, the majority fall somewhere in between. Consequently, such activities are classified as "provisional" because they can be undertaken if certain steps designed to minimize adverse impacts are followed.

Figure 8.3 schematically represents the land-use matrix concept. Although it is an oversimplification, it nevertheless demonstrates how some types of activities can be incompatible in one zone (e.g., excavation for fill in the wetland zone) while provisional or compatible in another (e.g., excavation for water management in the back-barrier zone). It is important to understand that the land-use matrix is not designed to restrict the use of private property, but rather to permit only those activities that are of low impact or that are designed to be compatible with recognized values in a given zone.

Permit Standards

The classification of certain activities as "provisional" requires the formulation of guidelines to be followed in the design and implementation of these activities. Such permit standards are written with one objective in mind: to guide "provisional" activities so that they will have a minimal impact on a zone's resource values. Because these standards cover a wide range of activities and seek to protect a wide variety of resource values and functions, this part of the resource management plan is the most difficult, yet most important, component of the Coastal Barrier Program. The permit standards must be prepared with the assistance and cooperation of local and state experts versed in coastal activities and barrier zone characteristics.

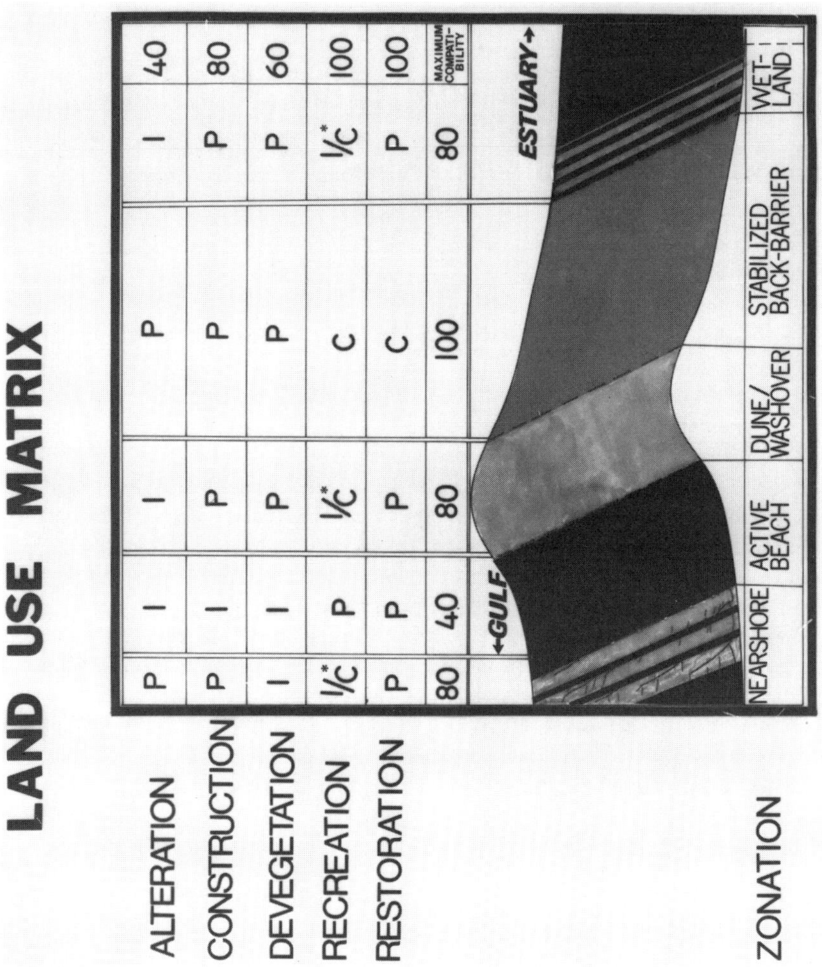

LAND USE MATRIX

	Nearshore	Active Beach	Dune/Washover	Stabilized Back-Barrier	Wetland	Maximum Compatibility
Alteration	P	–	–	P	–	40
Construction	P	–	P	P	P	80
Devegetation	–	–	P	P	P	60
Recreation	½C*	P	½C*	C	½C*	100
Restoration	P	P	P	C	P	100
Maximum Compatibility	80	40	80	100	80	

←GULF .... ESTUARY→

ZONATION

Figure 8.3. Land Use Matrix

Coastal Barrier Ordinance

A draft ordinance (Gore et al., 1984) has been developed for the protection and maintenance of coastal barrier resources in Collier County. It sets forth new administrative procedures that relate the level of permitting effort to the potential impact of a proposed activity. The major parts of this ordinance are:

(1) *Purpose and Intent.* This section states why the ordinance is being prepared. It lays out in a logical sequence the natural resource and community values of Collier County's coastal barriers. It describes the adverse natural and economic impacts associated with poorly planned coastal barrier development and the statutory authority for the Coastal Barrier Program.

(2) *Coastal Barrier and Coastal Mainland Protection Districts.* This section delineates the management areas to be covered by this ordinance. It defines coastal barrier and mainland zones within the districts and describes how to identify zone boundaries.

(3) *Administration.* These sections lay out the administration of the ordinance. They describe submission requirements and categorize proposed activities as compatible (no permit required), provisional (permit required from the Natural Resources Management Department), or incompatible (variance required from the Board of County Commissioners prior to commencement of the activity). These sections also outline the requirements for posting performance bonds and filing status reports.

Future drafts of the ordinance will also include permit standards and storm reconstruction procedures.

Summary

The coastal barrier system of Collier County is an important resource with both natural and community values. To ensure the protection and maintenance of these resource values, a Coastal Barrier Program is being developed. This program incorporates the information generated by the resource inventory and historical analysis into a management framework. The program identifies discrete coastal barrier units and beach segments within the county and highlights the coastal barrier features to be protected under the management program. Following the identification and classification of all coastal barrier activities, the program employs land-use matrices to permit reasonable land uses. At the same time, it protects recognized resource values by formulating permit standards to ensure that all coastal barrier activities will be designed and carried out with minimal impact.

The Coastal Barrier Management Program is being undertaken to protect the entire coastal barrier system by ensuring that only activities designed to minimize environmental impact are permitted. It is a dynamic program that can be amended as additional data are obtained. The primary goal of the program is to protect and maintain the resources of the coastal barrier system for their functional and recreational values, because coastal barrier protection is in the interest of all county citizens. It is hoped that with the aid of local community leaders, scientists, and planners, past mistakes in coastal barrier management can be avoided in the future.

References

Benedict, M.A., R.H. Gore, J.W. Harvey, and M.E. Curran. 1984a. *Natural Resources Management Plan.* Technical Report 84-1. Naples, Fla.: Collier County Natural Resources Management Department.
Benedict, M.A., R.H. Gore, J.W. Harvey, and M.E. Curran. 1984b. *Coastal Zone Management Units: Data Inventory and Analysis.* Technical Report 84-1. Naples, Fla.: Collier County Natural Resources Management Department.
Benedict, M.A., R.H. Gore, J.W. Harvey, and M.E. Curran. 1984c. *Coastal Zone Management Units: Atlas.* Technical Report 84-5. Naples, Fla.: Collier County Natural Resources Management Department.
Brown, M.T., and E.M. Starnes. 1983. *A Wetlands Study of Seminole County.* Technical Report 41. Gainesville, Fla.: Center for Wetlands, University of Florida.
Godfrey, P.J., and M.A. Benedict. 1977. *Natural Resource Management Plan for Cape Cod National Seashore - Phase I.* Report no. 23. Amherst, Mass.: National Park Service Cooperative Research Unit, University of Massachusetts.
Gore, R.H., M.A. Benedict, J.W. Harvey, and M.E. Curran. 1984. *Draft Ordinances for the Protection of Coastal Ecosystems.* Technical Report 84-6. Naples, Fla: Collier County Natural Resources Management Department.
Harvey, J. "An Assessment of Beach Erosion and Outline of Management Alternatives: Longboat Key, Florida." Contract Report to the Town of Longboat Key, Fla., 1982.
Harvey, J.W., M.A. Benedict, M.E. Curran, and R.H. Gore. 1984. *Coastal Barrier Resources.* Technical Report 84-2. Naples, Fla.: Collier County Natural Resources Management Department.
Pilkey, D.H., Jr., W.J. Neal, O.H. Pilkey, Sr. 1978. *From Currituck to Calabash: Living with North Carolina's Barrier Islands.* Research Triangle Park, N. Carolina: North Carolina Science and Technology Research Center.

REGIONAL MULTIPLE USE, LOCAL SINGLE USE: A POTENTIAL MODEL FOR REGIONAL COASTAL BARRIER MANAGEMENT

Hans Neuhauser

The Georgia Conservancy
Savannah, Georgia

The continuum of various types of coastal barriers found along the Atlantic and Gulf coasts may conveniently be divided into three groups: developed, preserved, and other—those neither preserved nor developed. Developed barriers include Jones Beach, Virginia Beach, and Hilton Head Island; preserved barriers include Chincoteague National Wildlife Refuge, Cumberland Island National Seashore, and Apalachicola Bay National Estuarine Sanctuary. "Other Barriers" include those identified in the Coastal Barrier Resources Act (CBRA). While the status of both developed and protected coastal barriers is not likely to change more than occasionally, the future of the "other" group—referred to here as CBRA units for convenience—is in considerable doubt. The central reason for this uncertainty is the unresolved conflict between proponents of economic development and those who believe that traditional barrier development defies logic in the face of such hazards as erosion, washover, storm surges, and sea level rise.

The strength of preservationist sentiment is growing, due in part to the educational campaign associated with the passage of CBRA. But economic forces are stronger in many cases. The challenge then is to find some mechanism that will prevent inappropriate barrier development while the educational campaign in favor of retreat from the shoreline proceeds.

One such mechanism is found in CBRA, which eliminates federal economic development subsidies such as flood insurance, highway funds, and water and sewer grants. Another mechanism, and one that may be more effective than CBRA in the long run, is presented here as a preliminary model called "regional multiple use, local single use."

The model is most easily explained by example. The region is the Georgia coast, a relatively short stretch of coastline, extending about 100 miles from its northern boundary with South Carolina and its southern boundary with Florida. Within this 100 miles of coastline, there are between 13 and 18 barrier islands.

(The number varies because there are different ways of counting them.) Each of these islands is used for one primary purpose, which varies from island to island. There are ten different principal uses represented among the Georgia barrier islands: (1) day use beach recreation, (2) mineral reserve, (3) wilderness for wildlife, (4) wilderness for recreation, (5) scientific research, (6) game management, (7) bedroom community, (8) nature park, (9) destination resort for high income persons, and (10) destination resort for average income persons. For each island there are also secondary uses that are, for the most part, compatible with the primary use. For instance, on the island where wilderness recreation is the dominant use, secondary uses include historic preservation, natural area preservation, research, and public education. On another island, which is primarily a bedroom community for a nearby city, compatible secondary uses include resort facilities, light industry, shopping opportunities, and historic preservation.

While each of Georgia's islands is used for a single primary purpose, a regional perspective discloses a de facto multiple use system (see Appendix). While not exhausting all possible uses of coastal barriers, the Georgia Sea Islands collectively account for a broad diversity of uses. An important distinction must be made between multiple use, as traditionally practiced by such agencies as the U.S. Forest Service, and what is presented here as regional multiple use, local single use. Most multiple use applications are for a single unit of land. For example, a single forest unit may be managed for timber production, watershed protection, hunting, hiking, nature study, wildlife management, and the like. At a given time, one use such as timber harvesting may be dominant, but over time other uses are accommodated.

In the regional multiple use concept, a unit of land is devoted to one purpose, for example, residential development, and that use does not change over time. There are no opportunities for hunting or wilderness recreation on a developed barrier and painfully few opportunities for nature study. Likewise, a preserved island does not change in use to accommodate development.

Some ecologists (e.g., Odum, 1975, pp. 54-56) have suggested that ecosystems containing a diverse array of species and interrelation-ships tend to be more stable and less vulnerable to perturbations than ecosystems with fewer species and interrelationships. The production of food from a mixed vegetable garden, for example, is more likely to continue in the event of a corn blight, since everything but the corn will continue to grow and produce. A large monoculture cornfield hit by a corn blight will be wiped out. The ability of a diverse ecosystem to be resilient to change comes at a price, though, because the diverse ecosystem is generally less productive than a monoculture.

This relationship between diversity and stability in ecosystems may also exist in the regional multiple use system. By providing a variety of different uses, Georgia's coastal barriers can meet most of the different kinds of demands that are placed on them. Because Tybee Island offers a good, high-density beach recreation area, for example, there is less demand by people desiring such an experience to develop nearby Wassaw Island, a National Wildlife Refuge, for the same use. The existing diversity thus lends resilience to the current use patterns for Georgia's coastal barriers. There is much less demand by the public to change one island's traditional use pattern into something dramatically different.

A MODEL FOR BARRIER MANAGEMENT 109

But like the vegetable garden, the resiliency comes at a price: the "production" is not as high as it could be under different circumstances. For instance, tourist revenue is not as high because some islands are not developed as resorts. Conversely, the number of people who want to camp out in a coastal wilderness area far exceeds the available supply of such areas. People are thus asked to share a limited resource with others whose interests are different.

Georgia's multiple use, single use approach is not the product of a grand design such as a coastal zone management plan: the state does not have one. Nor is the approach articulated at a state, regional, or local level. Georgia's system is due instead to a combination of historical accidents, fortuitous events, and a few dedicated individuals and agencies.

It is proposed that this regional multiple use, local single use system be considered for other coastal regions. In evaluating the appropriateness to other locations of the Georgia approach, some key points to consider include regional boundaries, the need for a particular use, the suitability of each component, and long-range flexibility.

The boundaries of the region need not coincide with the boundaries of the state. The manager of a coastal barrier in northern New Jersey may consider the relevant region as extending from Cape May to Montauk Point. The boundaries do not need to be permanent. Depending on the nature of a proposed use, a manager of a barrier on the panhandle of Florida may consider a large region from Apalachee Bay to Cat Island, Mississippi, or a small region limited to the west coast of Florida. Such flexibility in boundaries argues against the need to have the boundaries legally declared.

Management decisions regarding coastal barrier use within the region should be influenced by the availability of resources and facilities elsewhere in the region. Thus, the real need for a particular use requires careful examination. One special interest group may want 100% of the available resources for their use, to the detriment or even exclusion of other uses. To promote their cause, the special interest group may inflate or even invent demand figures. Managers and decision-makers must be aware of this propensity and guard against the acceptance of fictitious demand figures. Care must also be taken to identify truly barrier-dependent needs. Collecting seashells or watching shorebirds or allowing sea turtles to nest are barrier-dependent uses. Sunbathing or playing in the surf appear to be barrier-dependent but may not be. Sunbathers, given a choice, often prefer to congregate around the margins of a swimming pool rather than lie on a sandy beach. A swimming pool with a wave generator may be a practical alternative to opening a part of a wildlife refuge for high-density recreation. Still other uses, such as golf courses and tank farms for natural gas storage, do not require barrier location, and serious consideration should be given to preventing their establishment on such vulnerable and limited areas.

Management feasibility of the regional multiple use system depends upon the suitability of each component for its intended purpose. For example, if one beach is to serve as the outlet or "steam valve" for high-density beach recreation, the quality of that beach must be sufficient to meet the need. The concept of a high-quality, high-density beach access area is difficult to define. Some important components are identified below, along with some of the questions that should be addressed in order to establish such an area.

(1) A high-quality, high-density beach access area must have enough beach space available to satisfy the demand, and it must be available during both high and low tides. The adjacent waters should be sufficiently unobstructed by jetties and groins to provide safe swimming. Also to be considered is the cost-effectiveness of maintaining a beach large enough to meet peak demands that may last for only a few hours of a holiday weekend each year.

(2) The beach must be accessible both physically and legally. People need to be able to get to the barrier and then to the beach. Dune-crossing structures should be installed to minimize impacts to the most fragile component of the sand-sharing system. Adequate parking should be provided, but not necessarily at the edge of the beach. Satellite parking and shuttle transportation to various parts of the beach can reduce local overcrowding by taking people to less-utilized parts of the beach.

(3) Overnight accommodations and other services should be adequate for the expected demand. Facilities must be available for all income levels, at least on a regional basis. Reliance on market forces may be inadequate since excessively high room rates will eliminate the participation of people on the lower end of the economic spectrum. Public subsidy may be needed to assure equitable access. That subsidy may take the form of beach restoration funds or the establishment of a state park as at Jekyll Island, where rates are kept artificially low. Subsidy may not always be necessary but it should be considered as an option.

The regional multiple use scheme must also have some degree of flexibility built in. It is arrogant to assume that present day decision-makers can accurately predict what the best use of a coastal barrier will be 50 years from now. Some decisions regarding usage should be deferred for later generations to make. To maintain flexibility and provide future options, more barriers should be kept in their natural condition. In other words, the regional multiple use system should be heavily weighted in favor of natural area protection rather than development. The reason is simple: once a barrier is extensively developed, it is unlikely to revert to a natural condition.

The above considerations are not all-encompassing. In designating an area for a certain use, one should consider such factors as the integrity of nearby natural areas and the protection of important functions. For example, the development of a marina as an amenity for a residential project may pollute adjacent waters to the point that oysters and clams are no longer safe to eat. This has occurred on Hilton Head Island and is likely to occur in the Apalachicola Bay if further development occurs on St. George Island. If an area is selected for development, care must be taken to protect or restore the sand-sharing system of sand dunes, beaches, and offshore sand bars. This can be accomplished through the use of dune zoning, construction of dune crossover structures, restoration of nearshore sediment transport systems, and, where possible, the phasing out of all permanent structures on the shoreline.

These factors, while complex, do not present overwhelming obstacles to implementing a regional multiple use, local single use management system. They do, however, underscore the need to move away from laissez faire and into the arena of sound management clearly supported by adequate data. The greatest utility of the model comes in providing a flexible and logically defensible management system within which the fate of the coastal barriers can be determined in the face of continued development pressures and in light of the hazards of barrier development caused by washovers, storm surges, and sea level rise.

References

Odum, E.P. 1975. *Ecology: The Link between the Natural and Social Sciences,* second edition. New York: Holt, Rinehart and Winston.

APPENDIX

Primary and some secondary uses of Georgia's coastal barriers. The islands are listed from north to south. Public access symbols: E = actively encouraged, NE = not encouraged but allowed, and P = prohibited except on beach.

Barrier	Public Access	Primary Use	Some Secondary Uses
Tybee Is.	E	Day use beach recreation	Bedroom commmunity, average income resort
Little Tybee Is.	P	Mineral reserve	Day use beach, wilderness for wildlife, future investment
Wassaw Is.	NE	Wilderness for wildlife	Day use beach, hiking, hunting, research, education
Ossabaw Is.	P	Game management	Research, education, hunting, conferences, future investment
St. Catherines	NE	Research	Wildlife management, historic preservation, camping, fishing, education
Blackbeard Is.	NE	Game management	Day use beach, hiking, fishing, hunting, research, wilderness for wildlife
Sapelo Is.	E	Game management	Research, education, camping, hunting, timber harvesting, game export, conferences, future investment
Wolf Is.	NE	Wilderness for wildlife	Day use beach, fishing
Little St. Simons Is.	E	Nature park	Wilderness recreation, inn, education, hunting, fishing, nature study
St. Simons Is.	E	Bedroom community	Moderate income resort, airport, light industry, golf, shopping centers, retirement homes, historic preservation
Sea Is.	E	High income resort	Retirement community, second homes, education
Jekyll Is.	E	Average income resort	Retirement community, convention center, bedroom community, historic preservation, golf, education
Little Cumberland	P	Nature park	Second homes, natural area preservation, research
Cumberland Is.	E	Wilderness for recreation	Natural and historic preservation, education, inn

MANAGEMENT OF ADJACENT LANDS: LESSONS FROM THE NATIONAL PARKS[1]

Michael Mantell and Christopher J. Duerksen

The Conservation Foundation
Washington, D.C.

Introduction

Natural forces and development pressures in the coastal zone make management of national seashores and recreation areas increasingly complex. At Cape Cod National Seashore, Indiana Dunes National Lakeshore, and Gateway National Recreation Area (New York), for example, attempts by the National Park Service (NPS) to withdraw from expensive beach replenishment projects have sparked controversy between NPS and beach users, property owners, and local business interests. A new sewage line proposed in Cape Hatteras could mean added development with new roads and services. As the Coastal Barrier Resources Act takes hold, development pressure will presumably increase in already developed areas, many of which are adjacent to lands managed by the National Park Service. Moreover, the political rather than ecological nature of virtually all park boundaries exacerbates tensions arising from the needs of the parks and their neighbors.

The NPS mandate to preserve park resources unimpaired is difficult enough to apply inside the parks; it is far more complicated when pressures on park resources originate outside park boundaries or on privately owned lands within park boundaries. While it may raise hackles to close an overused area inside a park to hikers or fishermen, no one doubts the Park Service's authority to do so. When NPS combats outside pressures, however, its base of support is far less solid.

In a few situations, federal law provides the national parks with special protection against external pressures. Notable examples include the Clean Air

---

[1] This material is largely adapted from: The Conservation Foundation, *National Parks for a New Generation: Visions, Realities, Prospects* (1985).

Act, as amended in 1977, which provides visibility protection for 48 national parks, and the National Environmental Policy Act (NEPA), which requires that a federally supported or permitted activity consider in its environmental impact statement adverse impacts on the environment.(1)

Occasionally, NPS can exert influence through its control of access routes, water sources, or other resources without which nearby development is discouraged. For example, it is authorized to sell water to the gateway community of Tusayan outside Grand Canyon's South Rim if the park has a surplus and it would not be detrimental to park resources. Often, however, a park superintendent concerned about outside pressures must rely on measures available to any landowner, such as persuasion and pleading. A superintendent may appear at local zoning board and planning commission hearings to comment on projects that could adversely affect a park. But park personnel are often leery of being tagged as whining, unfriendly neighbors.

While recognizing the threats posed to park resources from outside activities, NPS has rejected recent proposals by conservation groups to add to the formal extraterritorial powers of the Park Service. NPS argues that if parks are viewed as meddling neighbors with authority to thwart development outside their boundaries, then (in the words of former NPS Director Russell Dickenson) new powers might "produce an extraordinary backlash against the park system itself." (2)

Responding to Outside Pressures

Lack of data, a chronic problem in formulating management plans *within* parks, is more serious when a case must be made for remedial steps outside. In some situations, NPS may suspect that adverse changes are taking place, but it cannot document the origin and/or extent of the harm. Sound information on external activities and their impacts upon the park is critical in coping with outside pressures. Staff trained to provide analysis of such effects is also critical to improving the NPS role.

Additional management tools are also necessary to protect National Park System lands. Opportunities for effective action depend on the type and location of the activity causing pressure:

- Activities on federal land adjacent to parks
- Activities receiving federal funding or permits
- Activities falling exclusively under state or local jurisdiction
- Problems affecting an entire region in which a park is located

Activities on Federal Lands Adjacent to Parks

Some pressures on the parks originate on adjacent federal lands. For example, the Forest Service or Bureau of Land Management may authorize mineral extraction, timber harvesting, or energy development on land they manage, and these activities may be visible from a park or may create noise pollution within it. Although such activities on adjacent federal lands are, in a sense, "within the family," they are not necessarily easy to reconcile. Each agency—sister bureaus within the Department of the Interior as well as those in other departments—has its own statutory mandates and constituencies.

Some impressive examples may be cited of coordination between NPS and other public agencies that administer adjoining lands. Federal, state, and local agencies work cooperatively, for example, in combatting forest fires. Perhaps the most widely known coordinating group is the interagency Grizzly Bear Committee that reviews projects, conducts studies, and proposes plans to improve grizzly bear habitats surrounding Glacier and Yellowstone National Parks.

More commonly, cooperation comes in the more limited form of an opportunity for NPS to comment on proposed agency activities that may affect a park as required under various federal laws, most notably NEPA. Most importantly, NEPA requires the lead agencies to consider the impacts of their proposed activities on the environment (which includes affected park system units) and to explore less-damaging alternatives. However, other agencies usually cannot be forced to do what NPS recommends, or even to assign park needs a high priority in weighing them against other objectives. In practice, agencies sometimes fail to take park impacts into account. The Department of Energy, for example, in its original consideration of a nuclear waste disposal site, did not consider the presence of Canyonlands National Park, less than two miles away.(3) Similarly, initial studies by the U.S. Army Corps of Engineers for modification of its Bluestone Dam to produce more power did not address the potential effects in New River Gorge National River immediately downstream.(4)

To redress this imbalance, conservationists have repeatedly called for legislation to strengthen the voice of the Park Service over activities on adjacent federal lands. Not surprisingly, other federal agencies are less than enthusiastic about what they see as attempts to expand park boundaries de facto. What is needed is a process that will produce thorough consideration of both park needs and competing policies and should include incentives to work out mutually acceptable solutions. Possible models include the consistency provisions under the Coastal Zone Management Act and the Endangered Species Act. The latter establishes a negotiation process that prompts agencies to search for a consensus on reducing harm to endangered species. A provision in the Act calls for the drafting of a conservation plan that would allow the proposed activity to go forward in exchange for efforts to enhance the environment of the endangered species and to protect certain habitat areas.(5) A similar review of projects on federal lands adjacent to the parks could prompt both NPS and the conflicting agency to sit down together and peacefully resolve their differences.

Review of proposed projects—even the type of cooperative review mandated by the Endangered Species Act—does not totally meet the needs of the parks and their federal neighbors. Effective reconciliation of conflicting needs requires cooperative planning as well, so that goals may be reconciled before projects are proposed. In practice, most federal lands adjacent to the parks are already subject to formal planning requirements which, along with NEPA, entitle the Park Service to comment on proposed plans for adjacent lands.

To enhance NPS effectiveness and ease the task of other agencies in addressing park needs, the Park Service should, with funding from Congress, and in consultation with affected federal agencies, identify specific areas of critical concern outside park boundaries, e.g., that serve as important habitat or migration routes for park wildlife, or that provide scenic views. Moreover, NPS should propose development policies for the designated areas that would accommodate park needs. In addition, such legislation should require other federal agencies, in preparing their plans and activities for adjacent lands, to be consistent to the maximum extent practicable with any specific NPS policies established for an "area of critical park concern." Precedent for such a consistency approach can be found in the Coastal Zone Management Act, which requires certain federal actions to be consistent with state coastal zone management plans.(6)

Workable mechanisms are needed to control potentially adverse impacts of projects on adjacent federal lands. If the federal government cannot deal with pressures originating on its own, it can hardly expect more of state or local governments or the private sector.

Activities Receiving Federal Funding or Permits

A number of activities that affect the parks do not take place on federal land but do have special federal involvement in the form of federal permits or funding. For example, the U.S. Department of Housing and Urban Development may fund construction of a high-rise hotel near an historic monument under its Urban Development Action Grant Program. A federal water pollution permit might be needed for a project planned upstream from a national park. The Environmental Protection Agency may be funding a major sewerage installation which will service new development adjacent to a national seashore on a coastal barrier.

As in the case of projects on federal land, several existing laws require the agency involved in such a project to allow NPS to comment—NEPA is the major one. By carefully evaluating the impacts of projects it approves or finances, the federal government can significantly enhance park protection. Yet more effective measures are needed to ameliorate the impacts that federally funded or permitted projects may have on national parks.

The Coastal Barrier Resources Act of 1982 is an excellent model. That law provides for identification of remaining undeveloped coastal barriers. Any new federal financial assistance for development in such areas—for projects like bridges and causeways—is prohibited. Thus, the federal government no longer subsidizes development in these flood- and storm-prone areas that may be eligible for federal aid and insurance payments, should storm damage occur.(7) Similar legislation for areas bordering units of the National Park System makes sense. Why should the federal government continue to subsidize or authorize projects that threaten the quality of its national parks?

An essential step in controlling federally funded and permitted activities is to define the scope of the Park Service's jurisdiction by preparing plans designating critical habitats, migration routes, and scenic viewsheds important to the parks. Also, there must be a reconciliation process designed to balance park protection and competing interests. Again, the Coastal Zone Management Act and the Endangered Species Act provide useful precedents.

Extending NPS jurisdiction will undoubtedly meet with stiff resistance, even with the added attraction of saving federal dollars. An interim step might be to make park protection applicable only to projects receiving a federal dollar subsidy, and to continue to rely on NEPA and other review mechanisms when permits are involved. In the final analysis, if the federal government is not willing to prevent itself from damaging the parks, it will have little justification for asking other interests and governments to do more.

Activities on Nearby Private and Non-Federal Public Lands

Many pressures on parks originate on private property under the jurisdiction of state and local governments. Such activities have a variety of impacts, and the ability of NPS to do anything about them varies from case to case. When a billboard is erected just outside the entrance to a park, or an abutting residential subdivision is approved by a local zoning board, there is rarely any federal involvement, and the Park Service is usually left to rely on cooperation from state and local agencies and private firms to protect park values. Sometimes such cooperation is forthcoming—the well of good feeling toward our national parks is deep—but often other goals take precedence, to the possible detriment of park resources.

A particularly thorny issue concerns private lands that are within national park boundaries—some designated as "inholdings" and the far larger acreage in new parks which is slated to be acquired or protected in other ways. Cape Lookout National Seashore in North Carolina, for example, has some 1,150 acres of private land within its boundaries, while Cumberland Island National Seashore in Georgia has nearly 4,000 acres of inholdings. Many assume that these lands within designated boundaries have some special protected status, but they do not, unless the federal government has taken other steps to protect them, such as purchase of a scenic easement.

Federal courts, including the U.S. Supreme Court, recognize the power of federal land agencies to constrain the use of nearby property in certain circumstances. Statutes give NPS this power in a few situations. Redwood National Park is perhaps the best example, where NPS has the power to regulate adjacent lands, as a measure of last resort, if the activity on them threatens to harm park resources.(8) Also, like any landowners, NPS has the power to constrain activities by neighboring landowners that are a "nuisance," that is, those interfering with the use or enjoyment of the property.(9)

At the heart of such reluctance to allow NPS intervention are respect for private property rights and resistance to intrusion into what are generally perceived to be purely local affairs. These values, coupled with a lack of data on the in-park effects of adjacent development, can often result in resistance by neighboring governments and property owners when the Park Service raises its voice against a project.

To address these concerns as well as park needs, protective measures need to be tailor-made, accommodating the diversity of parks and their local jurisdictions. Opportunities as well as needs vary from unit to unit. A plan that might succeed in New Jersey based on zoning controls, where state and local governments and courts have often supported strong land-use regulations, might not work in other places where land-use controls are often viewed with skepticism. Tax incentives for preserving historical resources have been more successful in Oregon than in many other places.

Currently, the most promising opportunity to protect the parks against external threats from private lands while accommodating local interests and attitudes lies in diverse cooperative mechanisms involving NPS and park neighbors. These partnerships are needed to provide a forum where activities can be discussed, differences thrashed out, and consensus developed. Often, such partnerships may be more effective if they involve a formal organization; land trusts are perhaps the most important example. Park officials have helped form some local land trusts and have sometimes advised local coordinating councils and land-use agencies.

At other times, less formal, cooperative ventures may be sufficient, and government need not always be the originator. There is, in fact, a precedent for private initiative in establishing such cooperation. For example, through a series of transactions involving the Appalachian Trail Conference and its affiliates, a 66-acre parcel was recently acquired for a trail right-of-way, scenic protection, and additional public use, including a hostel. NPS should actively encourage its staff to cooperate in such partnerships and private initiatives.

There is also a need for experiments at a few specified parks in which the NPS takes the lead in establishing partnerships with its neighbors. Several parks that have good relations with local governments, or at least the prospect thereof, might be designated to serve as models of successful federal-private sector partnerships. Such an effort should ideally be backed by an effort to make cooperation a reality, with Congress designating formal park protection zones, within which special tools and resources would be made available. Within these zones, Congress might give the park service authority to fund technical assistance and grants to pay for local land-use planning. In addition, NPS should be authorized to accept land donations and easements within these zones.

NPS should establish formal, ongoing ties with local governments. Several parks, such as Cape Cod, Golden Gate, and Lowell, have already improved their relationships with surrounding governments (and other community interests) through the formation of advisory commissions or by voluntary coordinating bodies.(10)

Regional Activities Affecting Parks

Some activities that put pressure on the parks cannot be handled well by a local superintendent or by drawing a new protective perimeter around a park.

Long-range transport of air pollutants falls into this category, as do activities that divert or pollute the water resources of the parks.

In 1977, Congress recognized that air pollution problems were affecting the parks by amending the Clean Air Act to provide for visibility protection in 48 national parks and national monuments. Another section of the Clean Air Act establishes a national goal to remedy existing visibility impairment in Class I areas, such as parks. Problems such as the long-range transport of air pollutants cannot be effectively addressed by treating parks in isolation. Special attention to these areas, desirable as it is, must be supplemented by action to address problems of the region as a whole. One example is a recent proposal that wildlife resources of Yellowstone National Park be managed on a geographical scale much larger than the park itself, so as to encompass the entire habitat range of the species in question.(11)

Conclusion

Increasing awareness that has resulted in NPS adopting a "let nature manage more of nature" philosophy will be most severely tested on coastal barrier lands within the National Park System in the coming years. Wind and sea constantly reshape barrier beaches, yet the accessibility of these areas to the public makes them prime recreational sites. Moreover, many coastal barriers contain communities of residents and tourist industries and have been altered by jetties, large parking lots, and shore development that accelerate the natural forces of erosion. How the National Park Service, affected coastal barrier communities, and interested citizens respond to these pressures will determine whether these areas will remain the special places that they are today.

Much more can and should be done to protect the parks against pressures from external activities. NPS should gather better data on the impacts of what is going on around the parks. It must better document the results of effective remedial action. And it must also show more ingenuity and willingness to embrace a variety of approaches to the problems it faces. Each park must develop a strategy tailored to deal with its particular set of outside threats, and NPS regional and national officials must be willing to back up park superintendents in carrying out the strategy. But NPS cannot respond to these pressures unless the Secretary of the Interior, the Congress, and ultimately the public are fully committed to supporting these actions.

Notes

1. 42 U.S.C.A. Section 7491 (1983); 42 U.S.C.A. Sections 4321- 4347 (1983).
2. U.S. Congress, House Committee on Interior and Insular Affairs, Subcommittee on Public Lands and National Parks, *Public Land Management Policy: Hearings,* Part III, 97th Cong., 2d sess., 1182, p. 440.
3. Ibid.; Conservation Staff interview with Terri Martin, National Parks and Conservation Association, 1984.
4. U.S. Congress, *Public Land Management Policy: Hearings,* op. cit., Part III, p. 440; Conservation Foundation staff interview with David Reynolds, resource manager, New River Gorge NR, 1984.
5. 16 U.S.C.A., Section 1536 (Supp. 1985); Fred Bosselman, "New Mechanisms for Dissolving Disputes in Federal Environmental Law," *Urban Land,* July 1983, pp. 34-35. This recent amendment to the Endangered Species Act was based on the process which had been used to resolve a controversy over development of San Bruno Mountain in California, habitat for the endangered Missions Blue Butterfly.
6. 16 U.S.C.A., Section 1456 (1974).
7. 16 U.S.C.A., Section 3501-10 (Supp. 1985).
8. 16 U.S.C.A., Section 796 (Supp. 1985).
9. Joseph Sax, "Helpless Grants: The National Parks and the Regulation of Private Lands, "Michigan Law Review 75 (1976): 262-263, 265-266.
10. C.H.W. Foster, *The Cape Cod National Seashore: A Landmark Alliance.* Hanover and London: University Press of New England, 1985.
11. *Boston Globe. "Yellowstone's Defenders Worry about Land Surrounding Park."* June 9, 1985, p. 15.

"CAMPGROUND TOWNS" OF THE SOUTH CAROLINA GRAND STRAND

Robert L. Janiskee and Paul Lovingood

University of South Carolina

Introduction

South Carolina's Grand Strand is a narrow strip of coastline that begins at the North Carolina border and extends southwestward in a gently curving arc about 55 miles to Georgetown. All but a tiny fraction of the region is sandwiched between the shoreline and the Intracoastal Waterway, a freshwater divide offering access to inland areas only by means of four widely-separated highway bridges. Thus, while the Grand Strand is not technically a barrier island, it is like one in that it is narrow, low-lying, exposed to storms, and restricted in access to inland areas.

The Grand Strand is blessed with site and situational advantages that have been exploited to create one of the nation's premier coastal resort regions. The climate is sunny and almost as warm as that of subtropical Florida. Few rivers or streams drain into the Atlantic Ocean along this coast, affording an unbroken stretch of white sandy beaches that extends for many miles. The beaches and water are quite clean because the area lacks major harbors and industrial sources of pollution. The region is easily accessible from the urban centers of the eastern U.S. and Canada and is approximately seven fewer driving hours from northern areas than the nearest Florida beaches (fig. 11.1).

Two decades ago, the Grand Strand was merely a summer resort for Carolinians and a stopping place for Florida-bound northerners. In the late 1960s, a major resort complex began to emerge, with Myrtle Beach at its hub. After 15 or so years of remarkable growth and development, the Grand Strand is now a booming vacationland with a visitor-based industry comparing favorably with many of the nation's best-known resorts (Smith, 1984).

About 300,000 visitors crowd into the Grand Strand on a peak summer weekend (augmenting the region's permanent population of about 48,000). These

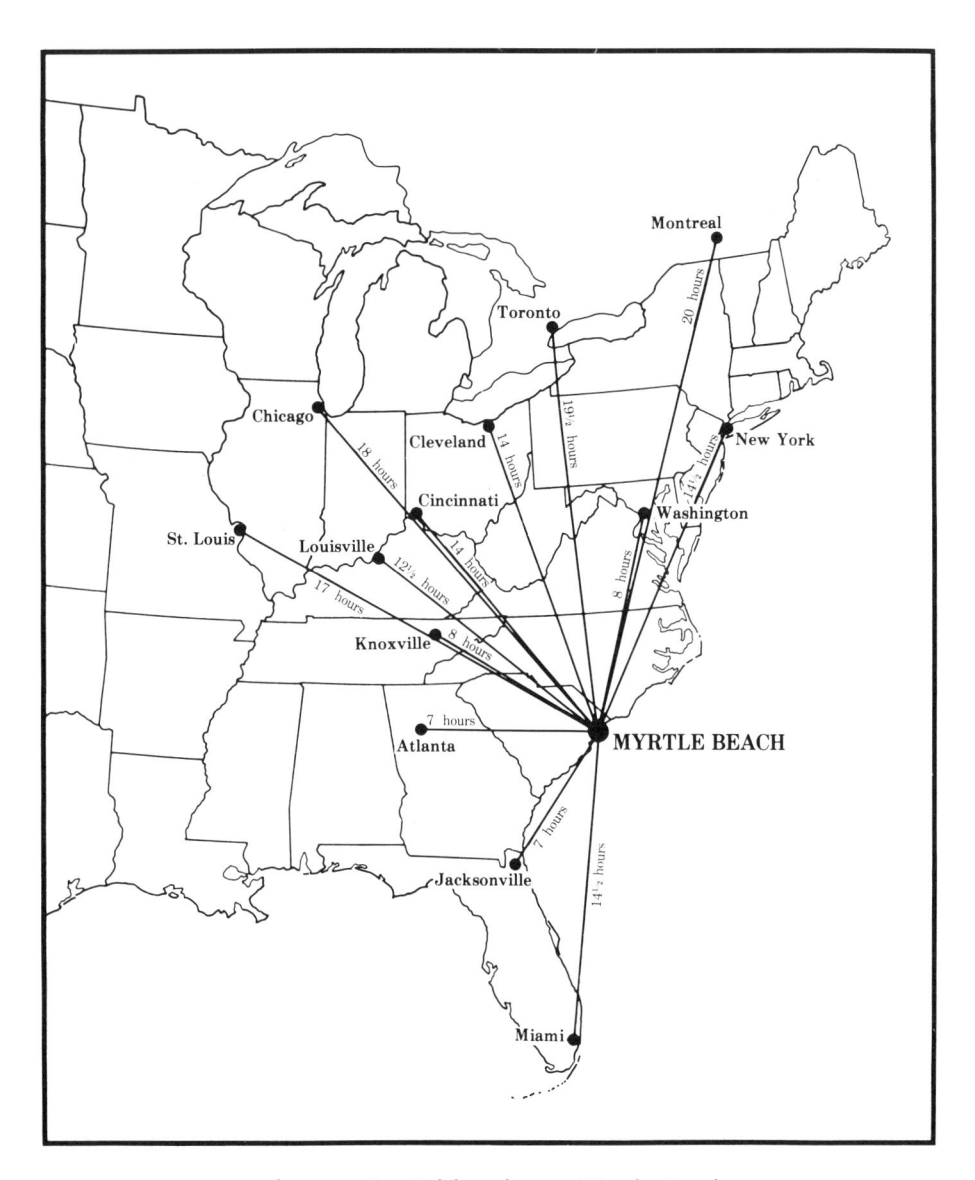

Figure 11.1. Driving time to Myrtle Beach

visitors are served by an impressive array of accommodations and amenities. An extraordinary feature of the Grand Strand's visitor-based industry is the strong contribution to the regional economy that is made by campers. The Myrtle Beach Area Chamber of Commerce estimates that campers spend about 20% of all tourism dollars brought into the region (Ward, 1982). This translates into roughly $200 million per year.

Myrtle Beach dubs itself "The Camping Capital of the World"—and legitimately so, since nowhere else does recreational camping take place at such a grand scale. During the busy summer season, the Grand Strand's ten privately owned and two public campgrounds accommodate over 30,000 people per day on about 10,590 campsites (fig. 11.2).

This regional camping complex includes several different types of campgrounds (Janiskee and Lovingood, 1982). The two public campgrounds, with an aggregate 419 campsites, are located in state parks. The most striking feature of the Grand Strand camping complex is an agglomeration of eight large, well-equipped commercial campgrounds with a total of nearly 10,000 campsites. Most camping in the Grand Strand occurs here. Most are oceanfront operations of extraordinarily large size (table 11.1). The big commercial campgrounds of the Grand Strand are so urban-like that we have dubbed them "campground towns" (Janiskee and Lovingood, 1983). They present some unique problems for infrastructure and hazard management.

Developmental History

The campground towns have an interesting developmental history. All eight were established in a flurry of development spanning the period from 1959 to 1971. At this time large parcels of highly desirable oceanfront property could still be purchased or leased at a reasonable cost. Moreover, entrepreneurs choosing to develop campgrounds on these parcels could do so unfettered by costly restrictions on land-use practices. Since a large segment of the camper market found the Grand Strand very appealing, developing campgrounds on oceanfront property in the Myrtle Beach area proved to be a lucrative form of investment.

By the mid-1970s, however, the development of additional large-scale campgrounds on the Grand Strand was foreclosed by the scarcity of affordable land. Oceanfront lots that had sold for $25 apiece around the turn of the century, and which had subsequently changed hands for some hundreds of dollars per acre, were now priced by the front-foot in thousand-dollar increments. Developers with enough capital to acquire the land needed for big campgrounds were now building motels and condominiums.

An additional constraint on new campground development had emerged in the form of federal, state, and local laws and regulations affecting the construction of camping facilities and the provision of utilities and vital services, such as potable water, electricity, sanitation, and security. Some of the practices that were

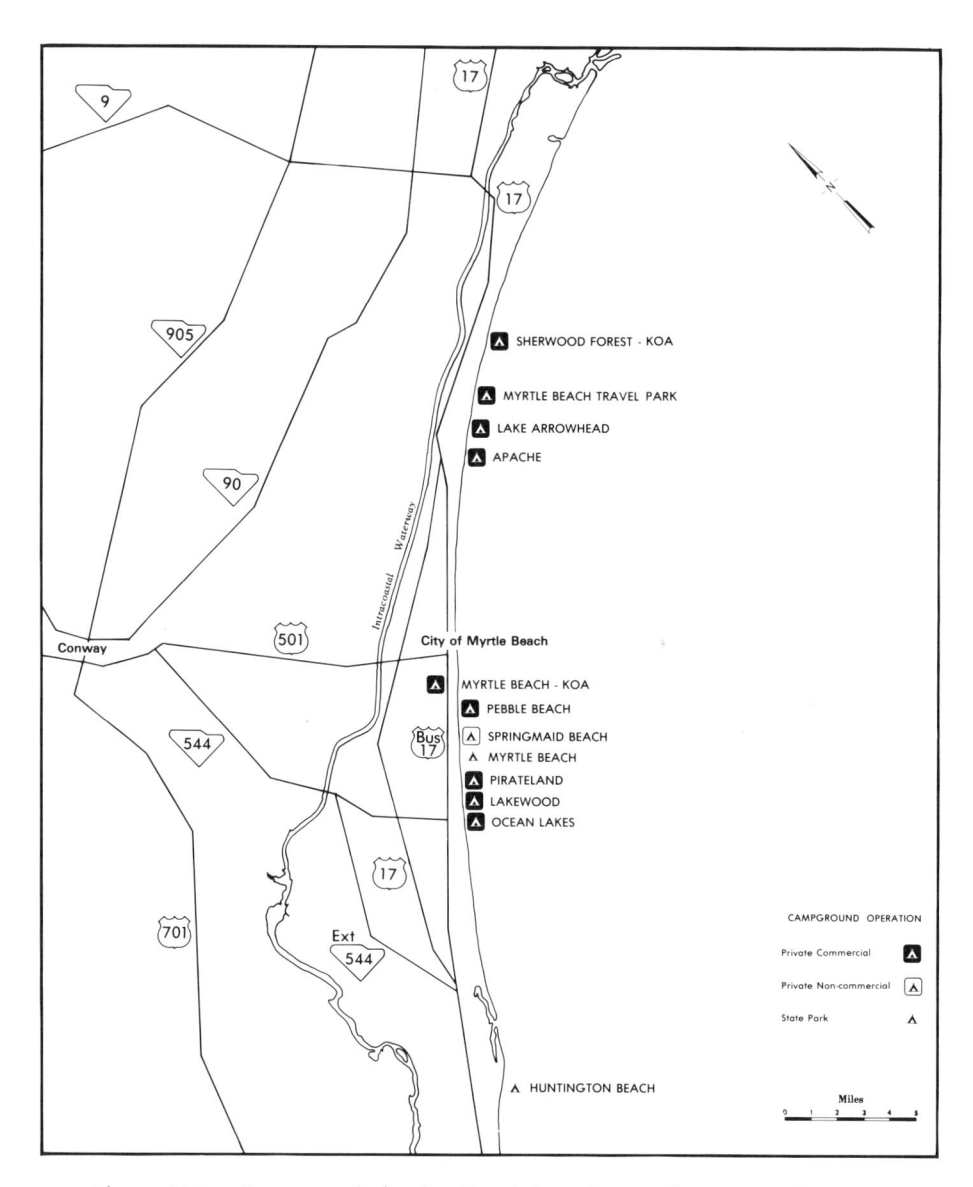

Figure 11.2. Campgrounds in the Grand Strand coastal resort region, 1984

TABLE 11.1

GRAND STRAND CAMPGROUND TOWNS, 1983

Campground	Approximate Number of Campsites*
Ocean Lakes	3,147
Lakewood	1,841
Lake Arrowhead	911
Pirateland	1,114
Myrtle Beach Travel Park	1,189
Apache	922
Sherwood Forest KOA	576
Myrtle Beach KOA**	330
Total Campsites	10,030

*With the exception of Pirateland's total, which was obtained from an advertising brochure, these data were compiled from reports furnished by the campground managers.

**Myrtle Beach KOA Kampground does not front on the ocean. All of the others have oceanfront locations.

used in the 1960s and early 1970s to construct camping facilities at a relatively low unit cost—leveling dunes, for example—were no longer permitted.

Existing campground towns are now under pressure to convert to more intensive development, such as condominiums. Several of the campgrounds are vulnerable in the short- or long-term future to the nonrenewal of land leases on which they are vitally dependent. Some are family-run operations that may become vulnerable to sale and conversion upon the retirement or death of key individuals. For these and other reasons, one can speculate that the heyday of the Grand Strand campground towns is already past and that they are destined to disappear by gradual attrition. But for the present time, at least, they remain a key component of the Grand Strand resort complex.

Management

Modern camping can be a highly developed form of recreation. The typical Grand Strand camper is not a person who cannot afford to stay in a motel; rather, he is one who chooses not to. Accordingly, a campground town is much more than simply a place to stay while enjoying the beach. It is a complex array of facilities and services catering to the special needs and tastes of a particular clientele. It caters largely to families rather than to singles or unrelated groups.

Camping in developed campgrounds has become a form of recreation dominated by participants who place a high premium on comfort and convenience. Today's campers are inclined to use more sophisticated forms of portable shelter, such as motor homes, travel trailers, camping (tent) trailers, truck campers, converted buses, and vans (Cole and LaPage, 1980, pp. 168-169). A large fraction of these recreational vehicles (camping RVs) are self-contained, having shower and toilet facilities, a stove, a refrigerator, and the proverbial kitchen sink complete with running water. Many are air-conditioned and some even have microwave ovens.

Camping RVs generally require campsites that are better equipped and have fewer obstructions than those designed for use by tent campers. Thus, a growing share of the campsites available in commercial campgrounds are "pull-throughs" with "full hookups" (Fuller, Foght, and Profaizer, 1980, pp. 183-184). Full hookups include water, electricity, and sewer connections, plus cable TV in some cases.

Many campers take out an annually renewable lease on a specific campsite. Leased sites, commonly termed "permanents" or "annuals," have become a conspicuous feature of many campgrounds, both large and small. The commonplace managerial practice of clustering permanents in particular parts of campground towns has created "neighborhoods" with an air of permanence about them. These neighborhoods are sometimes strikingly urban-like in appearance, especially when numerous sites are occupied by large travel trailers fitted with weather skirts and wooden decks. Since camping has traditionally been regarded as a transient activity, this raises the question of just where it is that camping ends and something else begins. This distinction appears not to trouble campground town operators or campers themselves.

Catering to the demand for leased sites offers several managerial advantages and entails at least one potential drawback. Campground areas that are unattractive to transient campers and plagued with below-average campsite occupancy rates can, in many cases, be successfully converted to leased-site neighborhoods. Regardless of where they may be located, leased sites provide cash flow in the form of annual payments that can be arranged to fall due in the off-season, when income from rentals to transients is at a low ebb. These and other advantages aside, there is a threshold level beyond which the further conversion of transient sites to leased ones in a particular campground results in an unacceptable erosion of the cash flow from transient rentals.

RV storage is another interesting facet of modern campground operations. Since the early 1970s, campers in growing numbers have ceased driving or towing their camping RVs back and forth to campgrounds. Instead they rent storage space in a favored campground and phone ahead to have their RVs moved to a site and readied for use upon their arrival. The appeal of this option is rooted in economic pragmatism as well as the quest for convenience. Increased fuel prices have dissuaded many people from driving their camping RVs any further than absolutely necessary. In addition, the downsizing trend in automobile manufacturing has yielded family autos that are commonly too small and underpowered to handle towable camping RVs. The resulting demand for storage services has been so great that a large campground may have anywhere from several hundred to several thousand camping RVs in storage. Storage is a lucrative enterprise, since RVs individually consume very little space and can be crowded into areas unsuited for campsites.

Campground towns are often destinations rather than stops on the way to somewhere else. A campground that functions principally as a destination-type facility can operate a sophisticated site-reservation system that books sites (or site guarantees) as much as a year in advance. During the peak season, especially in popular tourist locales, campers desiring guaranteed sites can be required to pay for a minimum stay of from several days to a week or more. This reduces campsite turnover, cuts overhead expenses, and tends to insure that the average camper will stay long enough to need or want a variety of camping-related services and goods.

Campground Towns and Coastal Storm Risks

The campground towns of the Grand Strand pose challenging problems for agencies concerned with land-use regulation and emergency preparedness. The Grand Strand region is physically analogous to a barrier island in macroform. While none of the region's campground towns is located on a barrier island per se, all are situated on a narrow and elongated coastal landform that is low-lying, exposed to storms, and bordered on the landward side by a water barrier, the Intracoastal Waterway. The Horry County communities in which the campground towns are located—Myrtle Beach, North Myrtle Beach, and Surfside Beach—were ruled eligible (during 1976-1977) to participate in the National Flood Insurance Program (NFIP). (The authors have not yet compiled information regarding NFIP-subsidized insurance coverage for the campgrounds per se.)

The Grand Strand's campground town complex is a potentially hazardous environment in which very large numbers of lightly-sheltered people and a similarly large stock of valuable personal property are concentrated in extreme density. The South Carolina coast is sporadically subject to severe storms, including hurricanes. The June through October hurricane season encompasses the principal season for Grand Strand camping, which spans the period from Easter to Labor Day. On peak summer weekends, the aggregate population of the campground towns approaches 30,000.

The last major hurricane to strike the Myrtle Beach area was Hazel, on October 12, 1954. Its violent winds and 17-foot storm surge laid waste to what was then a scarcely developed and thinly populated beachfront. If a similar storm were to revisit the area during the peak season today, the property damage would be enormous. Ample advance warning would be required to prevent heavy loss of life and injury among the campground town occupants.

Camping RVs and tents provide grossly inadequte protection against hurricane-force winds and storm surge flooding. On the Grand Strand, they are situated on low-lying ground only a short distance from the beach—many of them on sites containing large shade trees that can be uprooted with crushing force. Torrential rains and a storm surge of even modest proportions could convert thousands of camping RVs on the campsites and in the storage yards into floating debris. The battering-ram effects of this flotsam would compound the destructive power of the wind and water. A campground town is obviously no place for campers or their equipment during a hurricane.

It is also disconcerting to consider what might possibly happen if an evacuation of the campground towns were attempted in the face of a Category 3 or larger hurricane. Because a hurricane can intensify very quickly and/or veer ashore suddenly, a general evacuation of the Grand Strand beachfront might have to occur on very short notice—perhaps no more than 12 hours. Since certain areas of the Grand Strand are above storm surge level, it would not be absolutely necessary to evacuate all campers to the landward side of the Intracoastal Waterway. However, it is likely that most campers would attempt to drive or tow their RVs to safety. Traffic snarls developing in and near the campgrounds could become major arterial "gridlocks," blocking all movement in much of the region.

Problems similar to those we have described could also be produced by weather conditions less severe than those accompanying a major hurricane. Tropical storms, mature extratropical cyclones, and squall line thunderstorms are all capable of producing gale force winds and torrential rains along the Grand Strand beaches. Being unusually vulnerable to these hazards, the oceanfronting campground towns could sustain considerable damage from storms that cause little appreciable damage to conventional lodging facilities.

Fortunately, experience has shown that evacuating these campgrounds is feasible during non-peak periods, when occupancy is well below capacity. For example, the two separate evacuations that were made necessary by the erratic, threatening behavior of Hurricane Diana from September 10 to 12, 1984, were both accomplished quickly and uneventfully—even though the second one took place in the middle of the night. (No appreciable storm damage ensued, but it is encouraging that campground managers and campers took the threat seriously and did not hesitate to evacuate the campgrounds.) An evacuation taking place during the busy summer season might be similarly trouble-free, but the weight of greater numbers increases the likelihood of difficulty.

The Horry County Civil Defense Office, which is responsible for emergency planning and disaster preparedness, has in the past made no special provisions for evacuating campers (Bessent, 1985). However, the special problems posed by large populations of campers and camping RVs in the oceanfronting campground towns are being taken into account in the preparation of an updated evacuation plan for Horry County. Pending the implementation of this plan, campers will continue to be treated as members of the County's general transient population. It is standard policy to notify campground managers of the need for emergency evacuation and to provide them with relevant information such as that pertaining to evacuation routes and shelter availability. During episodes of unusually threatening weather, campground managers have almost always initiated inquiries to the county office without waiting to be notified.

Conclusion

Some of the general policies and procedures that have been developed for coping with the storm-related problems of seasonal or transient populations in coastal resort areas should be appliable to large-scale, destination-type campgrounds such as the campground towns of the South Carolina Grand Strand. Nevertheless, it is clear that campers are significantly more vulnerable to coastal storm hazards than beach visitors using conventional forms of lodging. Furthermore, the nature of the storm risks posed to people and property in campgrounds is sufficiently unique to warrant treating the campgrounds as a special category of coastal development in the contexts of storm risk assessment and emergency preparedness planning. In coastal areas where large populations of campers could be at risk, it seems prudent to assign a relatively high priority to the development and implementation of policies designed to minimize the likelihood of storm-related property damage and injury or loss of life associated with the use of camping facilities.

The coastal storm risk posed to camping facilities and equipment and to campers themselves is a multi-faceted problem with geographic and temporal variability. Accordingly, the measures developed to cope with this risk should be both comprehensive and flexible. The problems we have identified in this paper have implications for coastal land-use planning and regulation, campground design and management, emergency traffic control, emergency shelter siting, and other risk management measures, such as public education programs.

Suggestions for Future Research

Additional research, focusing on the storm hazard-related characteristics of oceanfront commercial campground operations, will be needed to place associated policy- and decision-making on a sound footing. Camper exposure to storm risks can also be reduced through increased public education and by improvements in campground design, campground management, and the safety-related features of camping equipment. Efforts should also be made to develop or improve measures for dealing with the potentially severe traffic problems that can arise when large numbers of camping RVs are evacuated from campgrounds.

Their storm hazard-related characteristics aside, oceanfront commercial campgrounds of large size represent a form of coastal urbanization that can have considerable ecological impact as well as significant implications for water supply, sewage disposal, recreational beach use, and other factors of importance to coastal barrier management. Further research effort might accordingly be directed toward identifying the specific nature of the problems characteristically associated with coastal urbanization in the campground town mode.

References

Bessent, W. Horry County (S.C.) Civil Defense Office, telephone interview, August 26, 1985.
Cole, G.L., and W. LaPage. 1980. "Camping and RV Travel Trends." In *Proceedings of the 1980 National Outdoor Recreation Trends Symposium,* vol. 1, pp. 165-177. Northeast Forest Experiment Station General Technical Report NE-57. Broomall, PA: Northeast Forest Experiment Station.
Fuller, C., P. Foght, and L. Profaizer. 1980. "Woodall Publishing Company, an Important Source of Camping Information." In *Proceedings of the 1980 National Outdoor Recreation Trends Symposium,* vol. 2, pp. 181-192, USDA Northeast Forest Experiment Station General Technical Report NE-57. Broomall, PA: Northeast Forest Experiment Station.
Janiskee, R.L., and P.E. Lovingood. "South Carolina's Grand Strand Campgrounds." Paper presented at the annual meeting of the SE Division, Association of American Geographers, Memphis, Tenn., 1982.
Janiskee, R.L., and P.E. Lovingood. "The 'Campground Towns' of the South Carolina Grand Strand." Paper presented at the annual meeting of the Association of American Geographers, Denver, Colo., 1983.
Smith, B. "Grand Strand Riding Big Wave of Success." *The State* (Columbia, S.C.), July 1, 1984, Sec. C, pp. 1,4.
Ward, A. Executive Vice President, Myrtle Beach Area Chamber of Commerce. Letter to Robert L. Janiskee, August 4, 1982.

ESTIMATING THE ECONOMIC BENEFITS OF BEACH RECREATION

Jonathan Silberman

University of Baltimore

Introduction

One of the public policy and management issues that developed coastal barriers face is the enhancement of beach access and recreation opportunities for the general public. Before any federal funds are allocated for a beach erosion control project, moreover, an estimate of the recreation benefits of such a project is required. There currently exists a substantial deficiency of information about beach recreation with which to make public policy decisions. The emphasis in studying developed coastal barriers is on technical rather than socioeconomic issues: geomorphology, erosion, ecological impacts, dune management, flood insurance, hurricane protection, and legal issues.

The information that exists on the economics of beach recreation is a by-product of erosion control studies conducted under the auspices of the U.S. Army Corps of Engineers (McConnell, 1977, and Bell, 1985, are the exceptions). Since 1981, there have been four beach erosion control projects evaluated by the Army Corps of Engineers on the East Coast: Virginia Beach, Virginia; Dade County, Florida; Sandy Hook to Barnegat Inlet, New Jersey; and Orchard Beach, New York. In each instance, recreation benefits were calculated. Earlier studies have been undertaken for Ocean City, Maryland and Myrtle Beach, South Carolina. The purpose of this paper is to identify and discuss some of the general issues involved in estimating the benefits of beach recreation. Detailed discussion of the techniques used is not included here. The discussion is divided into two general categories: existing and future beach usage.

Existing Beach Usage: Current Demand

The typical erosion control project often results in a widening of the existing beach. The initial approach to evaluating the recreation benefits of such a project is to determine the number of existing users and the recreation benefits of the additional beach to those users.

The first step in evaluating the recreation benefits of a beach nourishment project is to answer the question, "How many people currently use the beach?" Unfortunately, this question is difficult to answer. In most instances there is no systematic data collection of beach usage. Where localities charge a fee to park or to enter the beach and daily receipts are recorded, a fairly accurate estimate of beach usage is available. However, the information is not always precise because, in some cases, no record is kept of the number of season pass holders using the beach, or user estimates are derived from the number of parking receipts per carload.

In the absence of direct information, other sources are used. These include anecdotal information and state recreation surveys. Anecdotal information is derived from direct observation of beach usage, photographs, or estimates furnished by lifeguards and other safety officials based on, for example, automobiles in the parking lot. These estimates are similar to the "official" counts reported at demonstrations or political rallies. They are not accurate and tend to overestimate by a substantial amount. The inaccuracy is compounded because photographs and direct observations typically take place during the peak periods of beach use, holiday weekends. The beach usage observed at 1:00 p.m. on a July 4th weekend is not indicative of usage at other times of the summer season.

State surveys of recreation use are also not reliable. The surveys tend to be long and respondents are asked to recall their recreation activities in general rather than specific categories. For example, respondents in Virginia were asked the number of times they took part in swimming and sunbathing. Swimming and sunbathing do not take place exclusively at the beach. Even when specific questions are asked, respondents tend to overstate their recreation use. State surveys, moreover, do not account for visitors from outside the state, which can be a substantial number in the case of beach recreation.

In summary, there is a serious lack of information on the more basic form of beach recreation data, the number of people currently using the beach. The author strongly recommends the establishment of a systematic annual data collection effort to measure beach usage. Once quality data is made available, the underlying determinants of beach usage on a cross-sectional and time-series basis can be analyzed.

The establishment of a data collection system must incorporate the concept of time. Beach use should be defined in terms of a time dimension or flow. A reasonable interpretation of utilization would reject the extreme time span of 24 hours per day, 365 days per year, because most people use the beach for swimming and sunbathing. In February, for example, people will use the beach for walking, jogging, and relaxation. Utilization for swimming and sunbathing will restrict the time span to daylight hours and months with warm temperatures. Maximum beach usage occurs between 12:00 p.m. and 2:00 p.m. on summer weekends. Given the above discussion, the time span used to measure beach usage must involve an implicit judgment about utilization.

Due to the expense involved in measuring the flow of beach usage, sampling techniques and/or point estimates can be used. Furthermore, an origin survey that coincides with usage measurement would facilitate an understanding of the underlying determinants of beach use.

Existing Beach Usage: Willingness to Pay

Research in the area of outdoor recreation has supplied economists with tools for the economic evaluation of natural resources in recreational use. Although economists may disagree about some of the particulars of estimating recreational benefits, there is general consensus about what should be measured under ideal conditions. A general summary of the recreation benefit evaluation procedures has been published by the U.S. Water Resources Council (1983, pp. 77-87).

The procedure of preference is the contingent value method (CVM), which estimates consumer benefits from non-marketed goods through construction of a hypothetical market. "Willingness to pay" (WTP) is ascertained through a survey procedure which asks respondents how much they would pay to have the good provided rather than do without. Some of the earlier studies did this by asking for a definitive figure, i.e., "How much would you pay?" The superior method is to determine the WTP estimate through an iterative bidding process where the researcher asks, "Would you pay $X?" (For a discussion, see Randall, Ives, and Eastman, 1974.) If the response is affirmative, the interviewer raises the hypothetical fee by some arbitrarily chosen increment until the user will go no higher. If the initial response is negative, the interviewer lowers the hypothetical fee until agreement is reached.

In the interview process, income and other socioeconomic information can be gathered along with subjective variables, such as the perceived quality of the recreational facility. Number of trips per period, length of visit, years of experience with the facility, parking costs and availability, perceived availability and quality of substitutes, and so on can be determined. In addition, the subjective evaluation of the degree of congestion might be elicited. The interview process might also be accompanied by physical measurement of congestion during different time periods and on different days.

To analyze this survey data, the independent variable set is then regressed on "willingness to pay," the dependent variable. Information from the regression estimates can then be used to determine "total consumer surplus," given total use and characteristics of the user population. Consumer surplus represents the value of the quantity of the good purchased by the consumer, over and above the amount actually paid. Forecasts of expected consumer surplus values for future changes in the recreation facility (expansion, quality changes, etc.) or changes in user characteristics may also be made.

Despite rather widespread and growing use, the contingent value method is not without drawbacks or detractors. "Since the seminal article by Samuelson (1954), general agreement among economists suggests that any direct effort to value public goods will be plagued by the incentive structure facing individual consumers, encouraging them to misrepresent their true preferences" (Brookshire et al., 1980, p. 132). Fromm (1968) and many other economists believe the hypothetical questions asked in contingent (artificial) markets will generate invalid value estimates, because consumers believe that they would do just as well by bidding a zero or lower than true value and letting someone else pay. This is the classic argument as to why markets fail to provide the optimal quantity of public or quasi-public goods or, in the case analyzed here, non-marketed goods.

Certain more specific concerns about the value of information from contingent markets have arisen, including concerns that inappropriate valuations may result from incentives in the survey instrument. Responses to contingent valuation questions may elicit strategic behavior on the part of respondents. Respondents may understate the value of the good to them if they believe that the experimental results will be used to establish a price for the good. Respondents may also be influenced by the use of a hypothetical rather than a real market situation, by the method of payment (user fee versus higher taxes), and by the amount of information provided.

The author believes that the contingent value method is the appropriate technique for measuring willingness to pay for existing users. A good example is a survey developed by Mark Dunning and Dave Moesser of the Institute for Water Resources and used in Orchard Beach, New York in August of 1984. The approach taken by Dunning and Moesser was to extend the basic CVM to include a split design. Half the people interviewed were asked, "Considering today's weather and the number of people using the beach, would you pay $X to use Orchard Beach today?" The remaining respondents were asked, "Would you buy an admission pass to use the improved Orchard Beach shown in the picture if a daily pass costs $X per day?" The value of the incremental beach is determined by subtracting the WTP estimate for today from the WTP estimate from the improved beach. The samples above were further split to control for starting point bias. (The starting point is the initial valuation in a bidding game; this value may influence the final bid.)

It should be noted that the sample design for the CVM must match the concept of time over which beach attendance is derived. The results of the CVM will be biased if all the interviewing takes place during peak demand periods. It is anticipated that the net WTP for the improved portion of the beach will be higher during peak periods than during off-peak periods.

Other methods used to estimate WTP are the "unit day value" (UDV) and the "travel cost method" (TCM). The unit day value method requires the construction of a value per person per visit to a given recreation facility. The UDV for a particular project is determined by assigning points for the various recreational characteristics (U.S. Water Resources Council, 1983). The better and more varied the recreation experience, the higher the number of points would be. For example, a lake with fishing, swimming, and boating would receive more points than one with swimming only. The total points assigned are then converted into a dollar figure, which is the unit day value. Total benefits are a multiple of the UDV and the number of visits for a given time period.

The advantage of using the UDV stems from its simplicity of construction and minimum data requirements. All one needs to know is visitation and salient characteristics of the recreation facility. The drawback to using the UDV is that no relationship exists between the actual users and the benefit estimation. The value of recreation is determined by an outside "expert" ranking of the characteristics of the recreation facility. Everything is arbitrary and nothing about the process is in any way related to the actual behavior of users. In recognition of the highly mechanical and arbitrary nature of the UDV, the Water Resources Council disallows its use entirely on projects which will accommodate more than 500,000 additional visitors.

The travel cost method is based on a model for predicting use of a recreation site or area by visitors from surrounding origins. The general model can be described by an expression such as:

$$V_{ij} = (C_{ij}, S_{ij}, A_i);$$

where

V_{ij} = the number of site visits or trips per capita from a population source or center i to a recreation site j.

C_{ij} = trip cost, the cost of travel between the origin i and the site j, plus entry fees at site j.

S_{ij} = an index of the proximity of substitute recreation areas available to visitors from each origin i.

A_i = a vector of taste and socioeconomic characteristics of visitors from origin i.

Parameters of the model are estimated from information about users at existing sites. The model is then used to estimate expected visitations at a proposed site, provided that estimates of the relevant variables (C_{ij}, S_{ij}, A_i) are available. The same model can be used to estimate a demand curve for an existing or proposed site within the same region. The travel cost method estimates a demand function for a site or resource by using travel cost as a surrogate for price. The area under this demand curve provides an estimate of user consumer surplus or benefits.

The travel cost model is developed by using actual observations on use and user characteristics from various origins (i) to a site (j). The wide range of costs facing individuals at different distances from a site provides considerable information about the influence of costs on participation. This information can be used to generate a demand curve (i.e., estimates of participation at various entry fees).

TABLE 12.1

WTP ESTIMATES FROM RECENT
BEACH RECREATION STUDIES

Mean WTP Estimate	Location of Beach	Method Used
$ 3.26	Virginia Beach, VA	CVM - existing beach
5.31	Belmar, NJ	TCM - individual observations
1.53	Orchard Beach, NY	CVM - existing beach
1.61	Orchard Beach, NY	CVM - improved beach
2.50	Dade County, FL	UDV/TCM
1.42	State of Florida	CVM - existing beach (mail survey of local residents)

SOURCES: Bell and Leeworthy (1985); Silberman (1983); U.S. Army Corps of Engineers (1984, 1982).

The TCM has had extensive application for measuring recreation benefits over the years, because the method has certain very appealing qualities. In many cases the needed data are already available, having been gathered from visitor permits, surveys, and the like. Therefore, the method is often relatively inexpensive. In addition, there is little need for the time-consuming user contact necessary in some survey-based studies, such as contingent valuation, discussed below.

Despite the widespread use and acceptance of the TCM techniques, many conceptual and econometric difficulties have surfaced, including time costs, congestion, and multi-purpose visits. Some of these can be resolved, but many others cannot. The crucial conceptual difficulty is that the TCM can only be used to identify the value of the existing recreation resource. A change in a recreation resource, for example beach nourishment, cannot be evaluated using the TCM. Moreover, the TCM is an indirect rather than direct approach to measuring recreation benefits.

Presented in table 12.1 are the willingness to pay estimates derived for the four beach recreation projects referred to earlier. Substantial variation exists in both the WTP estimate and the technique used. Due to the small number of studies, it cannot be determined whether the results are sensitive to method and/or site characteristics. The author feels that an appropriately designed CVM is the most reliable method; however, it is the most expensive. For an in-depth discussion of the various techniques used to estimate WTP, see Desvousges, Smith and McGivney (1983) and McConnell (1983).

Future Beach Usage: Estimating Demand

The only way to forecast the future is to make use of historical information. Since accurate information on historical beach usage is nonexistent, forecasts of beach use are not likely to be reliable. Assuming that cross-sectional (point in time) information on beach usage and the origin (by zip code) of beach users is obtained, a forecast can be made using the travel cost method (Phillips and Silberman, 1985). A number of difficulties exist when using the cross-sectional approach. It should be emphasized that unless consistent, reliable estimates of attendance are collected on an annual basis, the only reasonable approach to forecasting is the cross-sectional one.

Two issues that require consideration when forecasting are definition of the study area and current non-users of the beach. In regard to the second issue, we have no information on whether or not people who currently do not use the beach would do so if the beach were restored. Surveying only beach users results in a sample selection bias.

When examining only current users, the project area beach can be considered in isolation. A forecast of future usage requires that the study area beach be placed in the context of a regional recreation resource. The geographic recreation "market" can be defined by the location of actual and potential user populations. The difficulty is that the restoration (expansion) of a beach may attract additional users from substitute recreation resources.

Consider, for example, the case of northern New Jersey. A project has been proposed to restore the beach from Sandy Hook to Asbury Park. The study area should encompass all beaches in the immediate region, which might include all the beaches south of Sea Bright. The boundary of the region, however, will be subject to judgment. Which boundary of the region, however, will be included and which should be excluded? Beach usage should be estimated and forecasted for all the beaches in the study area: Sea Bright, Long Branch, Asbury Park, Belmar, and Manasquon.

The problem now is to decide how many visitors to the regional beaches will alter their visitation and switch from their current beach to the restored beach at Sea Bright. The problem is made even more difficult when the dynamic aspects are considered. The decision to switch may be motivated, in part, by congestion at the existing beaches. If large numbers of people switch, the restored beach may then become congested, thereby altering the number who switch.

Two approaches to the above problem are suggested. The first is to survey beach users directly and ask them whether or not they would switch to the restored beach. Demographic and site characteristic data should also be gathered to develop models that help to explain the decision to visit one beach over another. The impact of congestion, attributes of the beach (parking, lifeguards, restrooms and showers, concessions, etc.), and travel cost will be crucial to an accurate forecast of attendance at the restored beach. The most straightforward approach to using the direct survey information would be to calculate the percentage of visitors from each beach that are willing to switch and apply this constant percentage to the individual beach forecasts. Essentially, the survey approach is a variant of the CVM. The above method is being used for the first time in a study of northern New Jersey beaches.

A second approach is to use an estimate of beach capacity and combine the concept of capacity with the forecast of recreational use. Once a beach (recreation resource) cannot accommodate any additional users (capacity), the excess demand will be attributable to the incremental or restored beach. This approach requires a definition of capacity; the problem is that any definition of capacity involves judgment rather than direct evidence. However, it is a relatively inexpensive method since no direct survey is required, such as in the first suggestion.

The crucial issues in defining beach capacity are the recreational resource standard and the time dimension over which capacity is measured. Assume, for explanatory purposes, that a beach has one million square feet. How many people can this beach accommodate? The beach recreation standard is the number of square feet per person designated necessary to provide a desirable recreation experience. No objective beach recreation standards exist; standards based exclusively on judgment range from 75 to 200 square feet. The number of people that a one-million-square-foot beach can accommodate at any one point in time will range from 5,000 to 13,333 people, depending upon the beach recreation standard.

Note that the capacity discussed is measured at one point in time. The flow of people on the beach is not considered. Evaluating the capacity of the beach over time is more realistic. As the time span increases, the capacity of the beach will increase. Each section of beach may be occupied by more than one person over time. It is important to consider what the most appropriate time span is, and whether or not weekday capacity should be separated from weekend capacity.

Conclusion

Beach erosion has reached critical proportions on developed coastal barriers, threatening the future of recreational services and storm damage protection. Federal, state, and local governments have recognized the need to intervene in behalf of beach preservation and restoration. While there seems to be adequate government authority, there is not adequate information on the value of beaches to aid officials in deciding the extent (in dollars) of funding for erosion control measures.

The benefits of beach erosion control can be divided into three general categories: (1) economic impacts on communities and businesses as a result of the expenditures of beach visitors, (2) the recreation benefits which accrue to beach visitors (consumer surplus), and (3) the storm protection benefits of a beach. Category 1 benefits are estimated using economic impact methodology. Category 2 benefits require estimation of a simulated demand curve using the travel cost method (TCM) or contingent valuation method (CVM). Category 3 benefits involve a detailed survey of all property near the ocean and probability calculations of storm damage. Category 2 and 3 benefits are those accepted by the Army Corps of Engineers.

This paper has briefly discussed the issues involved in estimating beach recreation benefits, an area of research which has been neglected. The lack of information on beach recreation may yield incorrect public policy decisions on erosion control projects such as seawall construction and beach nourishment. An understanding of the economics of beach recreation is crucial in formulating public policy for developed coastal barriers.

References

Bell, F., and V.R. Leeworthy. 1985. "An Economic Analysis of Saltwater Recreational Beaches in Florida, 1984." *Shore and Beach* 53, no. 2, pp. 16-21.

Brookshore, D.S., R. d'Arge, W.D. Schulze, and M.A. Thayer. 1980. "Experiments in Valuing Public Goods." *Advances in Applied Microeconomics.* Ed. V.K. Smith. Greenwich: JAI Press.

Desvousges, W.H., V.K. Smith, and M.P. McGivney. 1983. *Comparison of Alternative Approaches for Estimating Recreation and Related Benefits of Water Quality Improvements.* Washington: Environmental Protection Agency, Office of Policy Analysis. EPA Contract No. 68-01-5838.

Fromm, G. 1968. Comment in *Problems in Public Expenditure Analysis.* Ed. S.B. Chase, Jr. Washington: Brookings Institution, pp. 166-176.

McConnell, K.E. "The Economics of Outdoor Recreation." College Park: unpublished manuscript, University of Maryland, 1983.

McConnell, K.E. 1977. "Congestion and Willingness to Pay: A Study of Beach Use." *Land Economics* 53: 185-195.

Phillips, R., and J. Silberman. 1985. "Forecasting Recreation Demand: An Application of the Travel Cost Model." *Review of Regional Studies.* Section I of vol. 15, pp. 20-25.

Randall, A., B. Ives, and E. Eastman, 1974. "Bidding Games for Valuation of Aesthetic Environmental Improvements." *Journal of Environmental Economics and Management* 1: 132-149.

Samuelson, P.A. 1954. "The Pure Theory of Public Expenditure." *Review of Economics and Statistics* 36: 387-389.

Silberman, J. "Recreation Benefit Analysis for Virginia Beach, Virginia: Beach Erosion-Hurricane Protection Study." Norfolk: U.S. Army Corps of Engineers, Norfolk District, unpublished report, 1983.

U.S. Army Corps of Engineers. "Reanalysis of Federally Authorized Project, Atlantic Coast of New Jersey, Sandy Hook to Barnegat Inlet." New York: New York District, unpublished report, 1984.

U.S. Army Corps of Engineers. "Dade County Beach Erosion Control Study: Appendix 4 Economics." Jacksonville, Fla.: Jacksonville District, unpublished report, 1982.

U.S. Water Resources Council. 1983. *Economic and Environmental Principles and Guidelines for Water and Related Land Resources Implementation Studies.* Washington: U.S. Government Printing Office.

SHORELINE MANAGEMENT

APPROACHES TO COASTAL HAZARD ANALYSIS: OCEAN CITY, MARYLAND

Stephen P. Leatherman

University of Maryland

Introduction

The town of Ocean City, Maryland is located on Fenwick Island, an Atlantic coastal barrier, and it extends from the Delaware border south to Ocean City Inlet (fig. 13.1). Ocean City has been a resort community since the 1800s, but has experienced explosive growth during the last 15 years, characterized by the construction of many high-rise condominiums (fig. 13.2). The permanent population of Ocean City is still less than 6,000, but the transient summer population often exceeds 250,000 on peak weekends. It is now recognized that Ocean City is already highly developed with a tremendous economic investment in new real estate, that there are only limited opportunities for reducing the flood loss potential of this existing development, and that Ocean City will continue to receive strong pressures for continued development and redevelopment because of its established position as a major East Coast resort and its proximity to the major metropolitan areas of Washington, D.C. and Baltimore, Maryland (Humphries and Johnson, 1984).

Coastal barriers are dynamic landforms, subject to storm surge flooding and sand transport processes. These features are particularly vulnerable areas for human habitation since they extend seaward of the mainland and are composed entirely of loose sediment. Hazard planning for coastal resorts such as Ocean City often fails to recognize natural geologic and geomorphic processes and their consequences on the built environment and related habitation. In defense of planning methods, coastal hazard analysis often suffers from a lack of easily accessible, comprehensible data.

Efficient contingency plans require assembly of the following types of data: geomorphic analysis, shoreline and shoreface change analysis, and wave refraction analysis. A wave and storm surge analysis should also be undertaken, but this

144   COASTAL HAZARD ANALYSIS

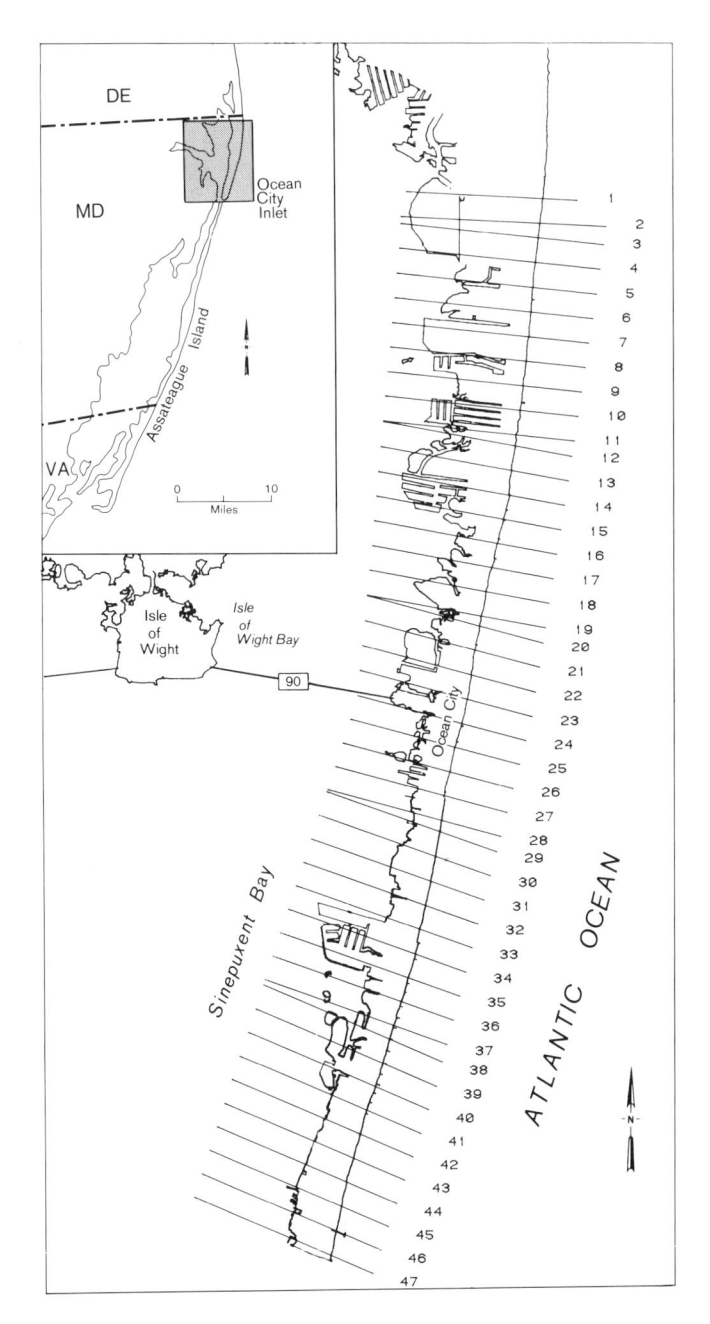

Figure 13.1. Location of study area along the Delmarva coast (inset) and index map showing transects used to depict shoreline changes

Figure 13.2. The manifestation of long-term recession of the shore is beach erosion, which becomes most apparent when comparing the soreline to fixed objects, such as buildings. Indeed, many of the high-rise buildings are already dangerously close to the water's edge.

subject has been adequately dealt with in the literature (Tsai, 1983; Jensen, 1983; Stone and Morgan, 1983). Finally, human modification of the barrier structure should be evaluated in terms of increasing or decreasing the hazard potential.

Geomorphic Analysis

Since coastal barriers are composed entirely of loose sediment—sands, gravels, clays, etc.—these coastal landforms are subject to erosion down through their entire core. While bedrock may be close to the surface on mainland areas, consolidated sediments often lie thousands of feet below the present barrier surface, far too deep to be of any importance in stability. A geomorphic evaluation of storm susceptibility involves a consideration of the nature of the three-dimensional stratigraphy of the barrier. Since sand is much more erodible than clay, inlets may preferentially breach through portions of the coastal barrier where the sand layer is thickest (Belknap and Kraft, 1985).

The Maryland Geological Survey has collected well-log and bore hole information which can be used to determine the depths to the sand-clay interface along the length of Fenwick Island. The areas with a generally deeper sand-clay interface can be inferred as being more likely to undergo the severe erosion and scouring that accompanies inlet formation (fig. 13.3). However, it should be noted

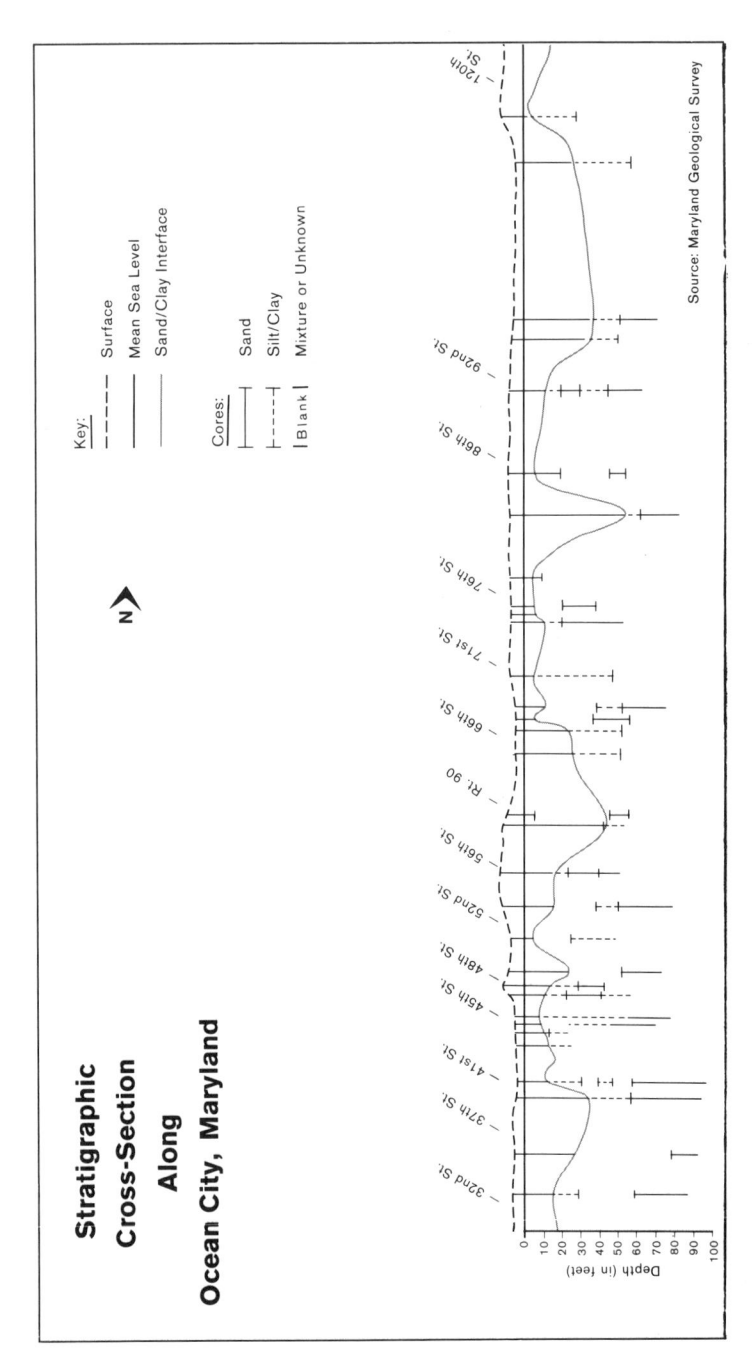

Figure 13.3. The 3-D stratigraphy of Ocean City, Maryland, shows an undulating subsurface along the length of this barrier, corresponding to now-buried Pleistocene fluvial channels of previous inlets. The depth to this compact clay under the sandy barrier core varies from 10 to 40 feet; those areas which are underlain by considerable thicknesses of just loose sand are the most susceptible to inlet formation when compared to adjacent sites with nearer surface contact of the more erosion-resistant clays.

that when this material is exposed on the shoreface by continued landward barrier migration, the fine-grained sediments (clay) will be largely lost offshore. Only the sandy materials have the potential to remain within the nearshore sand-sharing system.

Shoreline and Offshore Analysis

There is much historical information available for coastal resorts concerning shoreline movement trends, previous storms, and resultant damage. Qualitative historical data include old explorers' charts, noting the locations of past inlets and their migration, and written accounts of coastal storm damage. While this information can provide an indication of the severity of coastal storm conditions and the sudden geomorphic changes that can occur as a result, it is not accurate enough for historical shoreline comparisons.

Quantitative shoreline mapping depends upon the use of an existing data base from the U.S. Coast and Geodetic Survey (C & GS), now the National Ocean Service (NOS), that extends back to the 1850s for most coastal areas. Automated techniques for processing data and plotting maps from these historical sources, specifically using stereoplotters and metric mapping techniques, can be employed to obtain highly accurate information concerning shoreline trends (Leatherman, 1983a and 1983b). Figure 13.4 illustrates the historical shoreline changes for Ocean City, based on NOS data. The area has been subject to long-term, pervasive shoreline recession that has averaged 1.9 feet per year during the last 130 years.

Historical bathymetric comparisons can also provide important information; however, these data are not as readily available as shoreline movement information, and the inaccuracy is more questionable. A significant trend emerges from a historical bathymetric comparison of the area offshore from Ocean City (table 13.1). It is clear that the shoreface is steepening through time. The landward movement of the 20-foot depth contour is greater than the 10-foot depth contour, which in turn has migrated further than the mean high water line. The author conducted checks of the U.S. Army Corps of Engineers profiles used by Trident Engineering (1979), as compared to the original C & GS nautical charts, and obtained similar measurements (Leatherman, 1984).

It appears that the shoreline remains in approximately the same loction for a period of time, acting as a hinge as the adjacent shoreface steepens. While groins may have played a significant role in slowing down the rate of shoreline retreat at Ocean City since 1961, it is more likely that the lull in hurricane activity during the last 25 years is responsible for this trend. At the present time, it is not known what angle of shoreface inclination is the natural equilibrium orientation. The current, steepened condition offshore of Ocean City cannot be considered at equilibrium since recent bathymetric data have shown that the steepening trend has continued. Assuming that the equilibrium angle of inclination for the

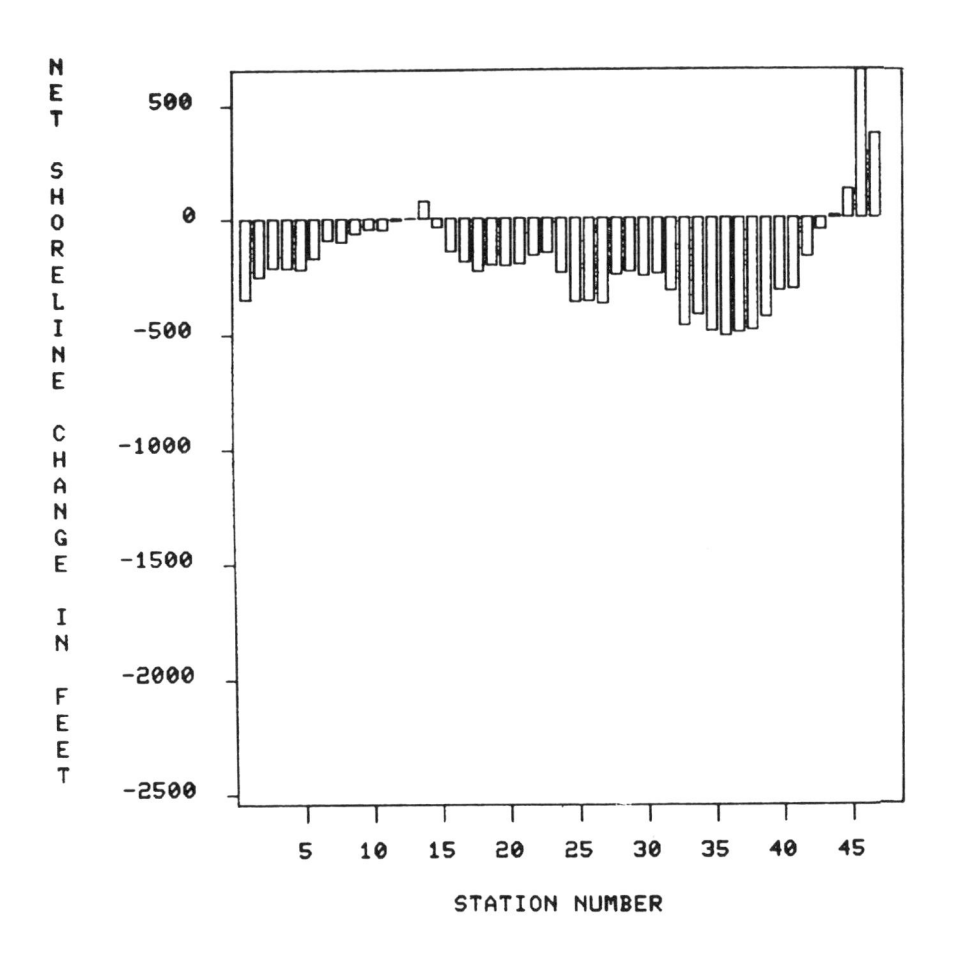

Figure 13.4. Histogram of historical shoreline changes (1850-1980) along the study area (see Figure 13.1). Positive values represent net accretion; erosion is indicated by negative values.

TABLE 13.1

CONTOUR SHIFTS (1929-1965) FROM
TRIDENT ENGINEERING
(1979)

	Over 36 Year Period	Average per year
Mean High Water (MHW) line	86 feet	2.4 feet
-10 foot contour	252 feet	7.0 feet
-20 foot contour	350 feet	9.7 feet

shoreface was reached at some point during the survey period (1850-1965), a future major coastal storm should cause the angle to decrease toward the idealized equilibrium position (Moody, 1964).

It is a well-established geologic principle that much geomorphic work is accomplished in quantum steps (Hayes, 1967; Leatherman, 1981). Therefore, a major coastal storm could provide the impetus by shifting and redistributing nearshore sands to reverse the steepening trend of the shoreface. At this point, the shoreface would return to its minimum angle and then continue to slowly steepen again until the next major storm.

In summary, the shoreface appears to undergo bicyclic adjustment through time. A long, quiescent steepening phase during which shoreline position is relatively stable or slowly retreating is followed by a brief, stormy period of shoreface flattening and rapid landward migration of the shoreline. Ongoing research should provide the data necessary to quantify this process and, hopefully, to formulate a predictability model.

Wave Refraction Analysis

Goldsmith et al. (1975) prepared a wave refraction analysis for Maryland and Virginia which predicts areas of wave concentration along the coastline for various storm conditions (table 13.2). The data indicate wave concentration in the northernmost portion of the study area, near the Maryland-Delaware border, and also a secondary locus at the Ocean City Inlet area. These data are not clearly reflected by the shoreline recession information, possibly because of the pulsaic nature of the barrier island retreat and downdrift migration of low-amplitude, very-long-period sand waves (Leatherman, 1984). As reported by the U.S. Army Corps of Engineers (1980), offshore dredging can also alter the characteristics of the incoming storm waves, perhaps causing abnormal concentrations of wave

TABLE 13.2

AREAS OF WAVE ENERGY CONCENTRATION
ALONG OCEAN CITY, MARYLAND*

Areas of wave refraction concentration	Number (Intensity)
120th street to Delaware Line	XXXXXX
60th St - 80th St	XXX
40th St - 60th St	XXXX
20th St - 40th St	XXXX
Inlet Area (up to N. Division St)	XXXX
100 St - 120th St	XXX
80th St - 100th St	X

*See figure 13.3 for street locations.

energies along certain stretches of the coastline. Therefore, erosion management on developed barrier shorelines should involve an analysis of the changes in alongshore wave energy concentration due to natural conditions as well as to offshore dredging.

Human Modifications

Human modification profoundly affects storm susceptibility of coastal barriers such as Ocean City. The built environment affects the barrier's ability to respond naturally to storm conditions in many ways. Among these are (1) interference with aeolian and overwash transport processes through paving and other construction, (2) dredging of salt marsh channels that provide greater likelihood of bayside inlet breaching during storms, (3) removal of the natural dune line, thereby increasing back-barrier susceptibility to storm waves, (4) construction of tall buildings which can force highly erosive currents through the gaps between them, and (5) lack of efficient storm water drainage, resulting in impounded water on streets, parking lots, and other paved areas.

On a coastal barrier that is in its natural state, sand is often transported from the beach to the back-barrier flats during storms. This transport occurs by both aeolian and overwash processes. During periods of quiescence following storms, some of this sand is transported back to the beach by the wind, aiding in natural beach restoration (Leatherman, 1980). Human intervention and development on coastal barriers such as Ocean City has diminished the effectiveness of this process. Sand is prevented from moving to the back-dune areas during storms by tall buildings and other structures which block and funnel the wind and water. These buildings may also diminish the occurrence of overwash, but more likely, will actually concentrate the flow into discrete areas between the buildings. Following a coastal storm, beach sand that has accumulated on streets and parking lots is removed by bulldozer and snow plow and is usually dumped on unused land, into the bay, or pushed back onto the beach. It is difficult to evaluate the short-term ramifications of this interference with the natural migrational processes, but clearly, the barrier surface is not allowed to increase in elevation as it would in a natural setting. Over the long term, the barrier could be drowned in place since landward migration is precluded and sea level continues to rise at an accelerated rate (Barth and Titus, 1984).

As the development of Ocean City progressed through the 1960s and 1970s, the amount of available oceanfront land quickly diminished. Real estate developers then looked to the bayshore as a potential area for construction of homes for people who enjoy boating. Since the bayside waters of Ocean City were quite shallow and often fringed by salt marshes, channels were dredged and the bay was deepened to allow for bayside dockage. The creation of finger-like channels perpendicular to the trend of the barrier greatly increases the susceptibility to inlet breaching at these locations in the event of a storm.

Coastal Flooding and Evacuation

Several hundred thousand people visit Ocean City on peak summer weekends, and the average summertime population is approximately 100,000. The National Hurricane Center normally gives six to eight hours warning concerning the likelihood of a hurricane (see Baker, this volume). Many coastal governmental authorities are reluctant to call for evacuations, fearing that if the danger does not materialize, people will become complacent and not take future warnings seriously. It is believed that residents and vacationers will tend to ignore evacuation warnings after a series of false alarms. As a result, many lives could be lost when a storm actually strikes. However, Ocean City has a unique problem that precludes the possibility of full-scale vehicular evacuation during a storm: the evacuation route is limited to only two bridges and Route 54 to the north. In the event of an evacuation warning, traffic jams and accidents would hopelessly snarl the way, leaving many thousands of people stranded in their cars or on the street, and in more danger than if they had stayed indoors. Rapidly rising storm waters, inefficient storm drains, and generally low barrier elevations could cause a high toll in lives, should an evacuation be necessary. Even if there were, miraculously, no accidents and minimal traffic jams, the likelihood of the bridges washing out or engines stalling due to inundation by rising waters still remains.

Clearly, vehicular evacuation is not the answer for a crowded resort area, although it is a viable option during the off-season. Vertical evacuation may be the safest way to protect people from hurricane-related deaths, 90% of which are due to drowning by the high storm surge. While the possibility of the undermining and toppling of high-rise buildings exists, such structural demise is not likely as long as building foundations are sufficiently deep, pilings are strong, and buildings are not located too close to the crashing surf.

Conclusions

Geologic and geomorphic data should be utilized in shoreline management, although a complete set of such data is rarely available and only utilized superficially in many instances. Geomorphic 3-D, historical shoreline/nearshore change, and wave/surge analysis indicate that certain areas are more prone to erosion or otherwise vulnerable during coastal storms. Human modification of coastal barriers such as dredging of bayside channels part-way through the island can also increase the likelihood of inlet breaching at these locations. With the recent improvements in scientific data gathering and analysis such as metric analysis of historical shoreline change, coastal planners and public administrators can incorporate this information for better management of development on coastal barriers.

Acknowledgments

This research was funded by the Maryland Department of Natural Resources and the U.S. Environmental Protection Agency. The assistance of Mark Eisner in report preparation is gratefully acknowledged. Susan Bresee is presently researching shoreface equilibrium conditions for her M.S. degree in geology at the University of Maryland.

References

Barth, M.C., and J.G. Titus, eds. 1984. *Greenhouse Effect and Sea Level Rise.* New York: Van Nostrand Reinhold Co.

Belknap, D.F., and J.C. Kraft. 1985. "Influence of Antecedent Geology on Stratigraphic Preservation and Evolution of Delaware's Barrier System." *Marine Geology* 63: 235-262.

Goldsmith, V., F. Byrne, A. Sallenger, and D. Drucker. 1975. "The Influence of Waves on the Origin and Development of the Offset Coastal Inlets of the Southern Delmarva Peninsula, Virginia." In *Estuarine Research.* Ed. L. Cronin. New York: Academic Press, Inc., pp. 183-200.

Hayes, M.O. 1967. "Hurricanes as Geological Agents, South Texas Coast." *American Association of Petroleum Geologists Bulletin* 51: 937-942.

Humphries, S.M., and L.R. Johnson. 1984. *Reducing the Flood Damage Potential in Ocean City, Maryland.* Annapolis: Maryland Department of Natural Resources.

Jensen, R. 1983. *Atlantic Coast Hindcast, Significant Wave Information.* Waterways WIS Report no. 9. Vicksburg, Miss.: Army Corps of Engineers Experiment Station.

Leatherman, S.P. 1980. "Barrier Island Management," In *Proceedings of Coastal Zone 80.* New York: ASCE, pp. 1470-1480.

Leatherman, S.P. 1983a. "Coastal Hazards Mapping on Barrier Islands." In *Preventing Coastal Flood Disasters.* Ed. J. Monday. Special Publication no. 7. Boulder, CO: University of Colorado Natural Hazards Research and Applictions Information Center, pp. 165-175.

Leatherman, S.P. 1983b. "Shoreline Mapping: A Comparison of Techniques." *Shore and Beach* 51: 28-33.

Leatherman, S.P. 1984. "Geomorphic Effects of Accelerated Sea-Level Rise on Ocean City, Maryland." Report to Environmental Protection Agency, Washington, D.C.

Leatherman, S.P., ed. 1981. *Overwash Processes,* Benchmark Papers in Geology. Stroudsburg, Penn.: Hutchinson Ross Publishing Company.

Moody, D. 1964. "Coastal Morphology and Processes in Relation to the Development of Submarine Sand Ridges off Bethany Beach, Delaware." Ph.D. dissertation, Johns Hopkins University.

Stone, G.W., and J.P. Morgan. 1983. "On the Concept of Storm Wave Susceptibility and Its Application to Four Florida Coatal Areas." In *Proceedings of 1983 ASBPA Meeting, Boca Raton, Florida,* pp. 19-36.

Trident Engineering. 1979. *Interim Beach Maintenance at Ocean City, Maryland.* Annapolis: Maryland Department of Natural Resources, Appendix B, pp. 1-14.

Tsai, Y.J. 1983. "Storm Surge Modelling in Coastal Floodplain Management." In

Proceedings of Coastal Zone 83. New York: ASCE, pp. 2005-2019.

U.S. Army Corps of Engineers. 1980. *Beach Erosion and Storm Protection, Atlantic Coast of Maryland and Assateague Island, Virginia.* Baltimore: Corps of Engineers.

DUNE MANAGEMENT: PLANNING FOR CHANGE

Norbert P. Psuty

Rutgers - The State University

of New Jersey

Introduction

There is growing awareness among the public and press that barrier islands are dynamic and migrating. Through the efforts of coastal geomorphologists actively involved in the public policy forum, state Coastal Zone Management (CZM) administrators and legislators have been made increasingly aware that past coastal management practices have not produced the desired objectives of reducing property damage, safeguarding public welfare, and maintaining the beach and other natural resources. Within the last two decades there has been a noticeable change in coastal protection policy, characterized by a shift away from the "quick fix" of groins and seawalls to the more "natural" management strategies of beach nourishment and protection of sand dune zones.

Dune Districts

In most coastal areas there is now some sort of dune protection program, albeit at a local level rather than as a uniform statewide practice. (See paper by Paul Godfrey in this volume.) State CZM programs extol the virtues of the "soft" or "natural" approach to shoreline management, such as the designation of dune districts as a zoning strategy for shoreline management in New Jersey, New York, and Florida (Gares et al., 1980; Nordstrom et al., 1978). The Dune Management District or Dune Zone concept is currently being applied by numerous communities to buffer erosion and to provide protection against storm surge penetration. "Keep off the dune" signs are now commonplace in public as well as private dune holdings.

During March 29-30, 1984, the Northeast Coast was struck by a severe northeasterly storm that caused considerable damage. The Governor of New Jersey proclaimed the state's coastline a disaster area and applied for federal relief funds. However, while lamenting the coastal destruction, he also stated that the state had been spared major coastal devastation because of the positive effects of the dune construction program applied during the past several years. The Governor's statement was also supportive of continuing efforts to use dune management as an intermediate-range planning tool.

While the recognition given to coastal dunes is much deserved, a paradoxical situation is now developing. The dune district was designed to accommodate natural processes and a gradual migration of the dune zone along with the rest of the coastal barrier (Gares et al., 1979). That is, in an erosional situation, the object of the dune district is to maintain protection and to buffer erosion during the gradual retreat of the beach profile. In current practice, however, the dune district is becoming yet another means to attempt to stabilize the shoreline. The dune line has become established on a map and has been followed by a variety of attempts to hold the dune and the dune line in place.

In some instances, it is apparent that the dune zone has been created by appearance rather than substance. That is, dunes have been planned and sited in locations where they cannot continue to exist. Locations of dune development in the upper portion of the active beach, or in front of buildings that are being undermined, are laudable but unrealistic. Sand fences, pilings, brush, Christmas trees, and other assorted debris placed seaward of the equilibrium position of dunes will not produce a viable dune system (fig. 14.1). Likewise, putting sand in place by bulldozer in nonequilibrium locations will not create permanent dunes.

At issue is a misconstrued interpretation of the dune district concept and the image of protection afforded by a coastal dune. The dune is being regarded and applied as an engineered structure in some areas. The dune zone is replacing the groin field as the attempted cure-all for shoreline erosion. Dune lines are being plotted on paper and built without regard for sediment budgets or displacement as sea level changes. Rock-cored dunes are being constructed to assure permanence of the feature only to assume later a seawall appearance and loss of the fronting beach.

Coastal dune zones are dynamic and integral components of the adjacent beach. It is as much a folly to attempt to isolate one from the other as to attempt to protect one as opposed to the other. The beach and dune comprise a sand-sharing system that interacts continuously. The understanding of this interplay is especially important for applying and maintaining a dune management program. The essential element is that *the dune zone may have to migrate if it is to retain its protective characteristics.*

DUNE MANAGEMENT

Figure 14.1. Sand fences positioned far seaward of equilibrium dune location. Fences and accumulated sand are likely to be removed during spring tides or storm tides.

Figure 14.2. Attempted dune enhancement by scraping sand from the beach to the face of the dune. The result is no change in total sand on the barrier profile.

158 DUNE MANAGEMENT

Application of the Dune District Concept at Fire Island

The following case study points to the dynamics of the beach/dune system and the variable positioning of the dune protection zone (Psuty, 1982). In 1976, the Fire Island National Seashore and the Townships of Brookhaven and Islip on Long Island, New York developed a dune protection ordinance. The ordinance was based on the establishment of a mapped dune crest line and a 40-foot buffer zone extending inland from the dune crest. An aerial photo mission was flown over the dune zone and stereo photos at scales of 1:2,400 and 1:1,200 were produced. By using a combination of the photos and field measurements, the position of the dune crest line was interpreted and transferred, along with the attendant 40-foot setback, to maps of the area.

The ordinance became effective five years later in 1981. At that time a second aerial flight was conducted. The same procedures for dune crest identification were followed, using 1981 aerial photos, and the two lines were superimposed on a base map using a zoom transfer scope. An arbitrary assumption was made that horizontal differences of 10 feet or less were modest and possibly within the margin of error. Thus, a -10 to +10 foot horizontal variation was defined as being stable.

The area of application on Fire Island consisted of private communities interspersed with the holdings of the National Seashore. Dune management was practiced along the entire length of the island. Many attempts at stabilization were already in effect, including sand fences, earth moving equipment, brush, and assorted debris barriers. The degree of control along the dune length varied according to public versus private ownership and also varied within specific holdings. Despite the manifold attempts at dune stabilization, only 45% of the dune crest line was within 10 feet of its original position after five years (table 14.1). Approximately 30% of the dune crest line had shifted inland or retreated, and 25% of the line had been displaced seaward.

Further scrutiny of the changes in the dune crest yielded information relating the displacement vector to the type of management being exercised. The dune crest categories were additionally subdivided into whether the dune was fronting an undeveloped area (usually a portion of the National Seashore) or lining one of the developed private communities. Field inspection showed that the park management practices were more benign, usually employing sand fences in breaches or in front of scarps. Private communities, on the other hand, practiced more rigorous methods of shifting sediment about with earth moving equipment (fig. 14.2) and locating sand fences and debris in the upper beach to encourage dune development.

All communities practiced some sort of dune stabilization; most were well intentioned but poorly based in their understanding of dune forming and stabilizing processes. Despite the laudable attempts at holding the dune crest position constant in these communities, only 48% of their dune crest length was classified as stable (table 14.2). Retreat (greater than 10 feet in five years) occurred along 33% of the dune crest length within developed areas, reaching a maximum

DUNE MANAGEMENT

TABLE 14.1

DUNE CREST DISPLACEMENT, FIRE ISLAND, 1976-1981

Stable (ft.)*	Erosional (ft.)*	Accretional (ft.)*
29,687	19,776	16,496

*Length of shoreline in category.

TABLE 14.2

DUNE CREST DISPLACEMENT, FIRE ISLAND, 1976-1981
IN PARK AREAS AND IN DEVELOPED COMMUNITIES

Stable (ft.)*		Erosional (ft.)*		Accretional (ft.)*	
Park	Develop.	Park	Develop.	Park	Develop.
12,754	16,933	8,171	11,605	10,028	6,468

*Length of shoreline in category.

TABLE 14.3

AREAL DISPLACEMENTS OF DUNE CREST LINES
FIRE ISLAND 1976-1981

Erosional (sq. ft.)		Accretional (sq. ft.)	
Park	Develop.	Park	Develop.
256,064	252,127	520,900	202,058

displacement of 43 feet in this period. The smallest proportional change, 18%, was recorded in that component of the developed shoreline where the dune crest line was displaced seaward by more than 10 feet.

The essential point is that the dune district concept was never intended to be a static phenomenon, and the measurements at Fire Island demonstrate this situation vividly. To cause the dune zone to stay in place would be to reintroduce the seawall syndrome—a bit more environmentally acceptable, perhaps—but a seawall nonetheless. Profile lines were surveyed in 27 locations to establish the vertical dimensions as well as the planimetric data available from the aerial photos. Nearly 25% of these profile lines recorded shifts of the dune crest line greater than 40 feet. Thus, in these very dynamic areas, the application of the dune crest line and its setback zone had been totally compromised by the time the dune ordinance went into effect.

The significance of the dune crest line changes is enhanced by further analyses of the situations representative of directional shifts. It is assumed that in those stable areas of less than 10 feet displacement, there is little gain or loss of the beach/dune system. However, the 30% of the total dune crest length that was displaced inland represents a total areal loss of over 500,000 sq. ft. (table 14.3). This value represents the total area of the polygons described by a comparison of 1976 and 1981 dune crest lines. An inland shift of the dune line would occur only when volume is removed from the seaward margin of the dune. Thus, the loss of dune area represents a loss of sediment from the face of the dune, despite attempts to hold the crest and the form in place.

Seaward displacement of the dune crest line could represent volumetric accumulation, but generally this was not the case. Rather, displacement was the product of either of the following scenarios:

1. A straightening of an irregular dune crest line wherein blowouts were crossed by sand fences that were subsequently mapped as the dune crest line
2. Fence lines established seaward of the previous dune line that had trapped a sufficient amount of sand to be considered the new dune crest.

In either of the above cases, the areal growth represented by the seaward displacement of the dune crest line was not a volumetric gain equal to the increased width of the dune. The seaward displacement took the form of an accumulation about the new fence line and thus the additional sediment in the system was very limited. It is likely that the accumulation was less than the total volumetric loss, representing a net loss.

Conclusions

Thus the beach/dune system at Fire Island shifts considerably, even while attempts to stabilize it are ongoing. Other communities and states using similar approaches to dune protection and management likewise fit into the same syndrome, namely, they are trying to use static measures to control and plan for a dynamic environment. Dune management is an acceptable land-use device only when a periodic review of the dune crest line is conducted and the position adjusted accordingly.

Dune management, dune districts, and dune zoning are viable strategies to use in shoreline management and planning, but they must be used in conjunction with present knowledge of the physical system. It is impossible to stabilize the dune zone in the face of a continually eroding beach. It is impossible to stabilize the dune zone in the face of a rising sea level. Each of these conditions will result in an inland displacement of the dune zone or its reduction and elimination. Coastal scientists have heightened public awareness of the value of coastal dunes in the beach/dune system but have made little progress in communicating the necessity of the mobility of the dune form as a consequence of deficit sediment supplies and a rising sea level. [See following paper by Godfrey for a case study of dune restoration, ed.]

References

Gares, P.A., K.F. Nordstrom, and N.P. Psuty. 1979. *Coastal Dunes: Their Function, Delineation, and Management.* Technical Report NJ/RU-DEP-14-12-79. New Brunswick, NJ: Center for Coastal and Environmental Studies, Rutgers University.

Gares, P.A., K.F. Nordstrom, and N.P. Psuty. 1980. "Delineation and Implementation of a Dune Management District." *Proceedings Coastal Zone 80.* New York: American Society of Civil Engineers, pp. 1269-1288.

Nordstrom, K.F., N.P. Psuty, and S.F. Fisher. 1978. "Empirical Models of Dune Formation as the Basis for Dune District Zoning." *Proceedings Coastal Zone 78.* New York: American Society of Civil Engineers, pp. 1489-1507.

Psuty, N.P. 1982. *Comparison of Dune Crest Line Position, 1976-1981, Fire Island, New York.* Technical Report PX-1750-1-0450. New Brunswick, NJ: Center for Coastal and Environmental Studies, Rutgers University.

A SUCCESSFUL LOCAL PROGRAM FOR PRESERVING AND MAINTAINING DUNES ON A DEVELOPED BARRIER ISLAND: MANTOLOKING, NEW JERSEY

Paul Jeffrey Godfrey

University of Massachusetts

Introduction

Efforts to stabilize beaches and dunes through federal and state programs are increasingly costly and fraught with political uncertainty. Accordingly, a few communities have taken the lead in developing their own programs to protect coastal dunes. One such community is Mantoloking, New Jersey, a heavily developed coastal barrier between Bay Head and Normandy Beach, north of Island Beach State Park.

Development of the Mantoloking Beach began in the late 1800s with major growth after World War II, particularly during the 1960s and 1970s. Some of the older homes were washed away as much as 70 years ago, with losses continuing until 1962. Since then, none has been lost. The barrier is developed from the bay side to the beach, with two major roads running the length of the town. Mantoloking is about two miles long and nearly every beachfront lot contains a house. The buildings are substantial and many are year-round residences. Many of the older homes remain, along with numerous more modern structures. There are currently 117 owners of beachfront property.

Nearly all the buildings in town still have a dune between them and the sea. Much of this dune was destroyed during the Ash Wednesday Storm of 1962, but it has been rebuilt by natural processes and human efforts. The beach has been relatively stable since then, with some storms washing up to the dune edge. However, human-caused dune deterioration was occurring at an increasing pace, to the alarm of residents who understood the need to preserve the dune. Efforts began in 1980 to develop a dune ordinance and to promote public awareness of the value of the dune system.

This brief report will describe the steps taken by the town in developing its successful dune policy. The program has three parts: a dune ordinance, monitoring of the dune system, and a public education effort.

Dune Ordinance

In 1980, the Borough Council adopted "An Ordinance to Regulate, Preserve, and Protect the Beaches and Dunes within the Borough of Mantoloking, Ocean County, New Jersey." The introduction recognizes the value of coastal dunes as a barrier against storm damage and the need to protect the dunes from human disturbance. Definitions that relate to major features of the program are listed.

The ordinance prevents any construction behind the defined setback line and requires that any new construction be above a specified height. It prohibits people from walking on the dunes and requires landowners to maintain only one narrow pathway across their dune. Dune platforms are not to exceed 200 square feet in area and must be built according to required specifications. The ordinance prohibits disturbing natural vegetation or removing sand from the beach or dunes. Motor vehicles may not be used on any dune area in town.

Landowners may not lower the dune in any way and must build up their dune if it is less than a recommended height. If an owner lowers his dune and refuses to rebuild it when so notified, the Borough may remedy the situation at the owner's expense. Landowners are required to plant dune vegetation, erect fences, or both, to restore and build up the dune system, according to the advice of the "Dune Consultant."

The ordinance established the position of Dune Inspector, with authority to enforce the regulations. Landowners who do not comply can be prosecuted by the Borough.

One might think that such a strong ordinance would meet with considerable hostility from landowners, but thanks to advance publicity and many meetings regarding the need to protect the dunes, there has been nearly complete compliance with the regulations. Once landowners realized that any effort to build and protect their own dunes could fail if all did not work together, there was a marked change in attitudes toward the program. Since 1980, there has been remarkable progress in dune restoration, fencing, grass plantings, and construction of wooden walkways over the dunes. The results of this effort were seen after storms during the winter of 1984. Parts of the New Jersey coast were devastated. Mantoloking's dunes were truncated, as shown in figure 15.1, but there were no major problems except where the beach was too narrow, or the dunes still too low.

There have been two challenges to the ordinance, one of which was resolved in court in favor of the Borough. One resident was lowering his dune by creating

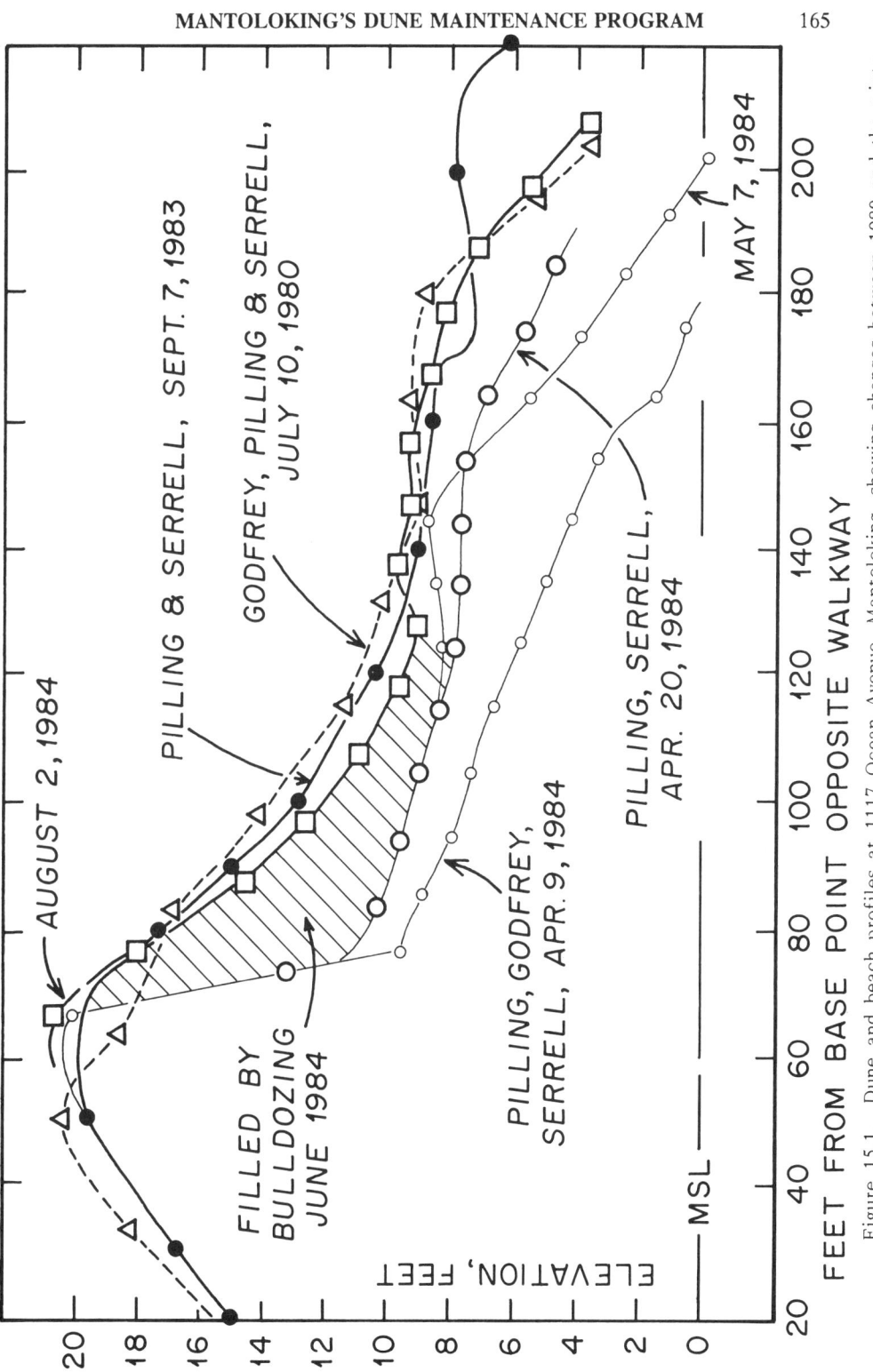

Figure 15.1. Dune and beach profiles at 1117 Ocean Avenue, Mantoloking, showing changes between 1980 and the winter storms of 1984 with subsequent recovery, both natural and by bulldozing. Plotted by J. J. Serrell and redrawn by P. Robinson

"channels" on either side of his property. He was told that this was a violation of the ordinance and was required to restore the dune. The landowner then sued the Borough and subsequently lost his case in court. The judge ruled that Mantoloking's Dune Ordinance served the public good and was not a violation of the owner's rights. The author served as an expert witness for the Borough, demonstrating the potential effects of dune breakthroughs and washovers. Since then, the channels have been filled and the dune restored. There have been no legal challenges since. Another landowner resisted repeated urging by the Dune Inspector and the Council to build up his dune. This has now been accomplished by the Borough at the owner's expense.

The ordinance also provided a program for the Borough's own dune access points. Prior to this program, physical dune crossings were inadequate. Many public ways were very low and provided easy passage for flood waters through the dunes. The Borough began a program to construct substantial boardwalks with steps down to the beach. The boardwalks have gaps between the planks so that light can penetrate, and they are high enough that dune plants can grow underneath and sand can accumulate. As the dune grows, the boardwalks can be raised. The program has been very successful and provides a model for other efforts.

The Borough also built a vehicle ramp for official use over one dune access point, using plans obtained from the National Park Service. The ramp was made by attaching two-by-four studs together with cable, leaving intervening spaces. Once in place, the ramp prevented the lowering of dune elevation as vehicles passed over. The ramp settled into the sand, but could easily be lifted and placed on new sand as it accumulated. Beach grass invaded the sand space between the studs and the plants survived, even though vehicles occasionally passed over. Their roots and rhizomes were protected from tire damage, the main cause of plant death due to vehicle passage (Godfrey and Godfrey, 1980).

The Borough also provided recommendations for constructing dune platforms. These structures are being observed regularly to determine whether they have any adverse effects on the dune. Those that are built above the sand, with gaps between the boards so that light can reach the plants below, seem to cause minimal problems.

Dune Inspections and Measurements

A second important part of the Borough's dune program is periodic monitoring of the dune system. At least twice a year (summer and winter), the Dune Consultant, Dune Inspector, Council President, Editor of the *Mantoloking Reporter*, and other interested persons walk the entire shoreline in town and inspect each landowner's dunes. Notes are made about the general condition of the dunes, the efforts at dune fencing and planting, type and adequacy of dune boardwalks and steps, effects of previous storms, and any needed improvements or notices. The Dune Inspector makes more surveys as needed and provides technical assistance to landowners.

Following the winter storms of 1984, the dune was cut back considerably, but not behind the dune crest in most areas. Soon after, the beach began growing seaward again, and sand was added naturally to the beach and dune face. Some landowners elected to speed up the restoration process by bulldozing sand to the dune face. Within a very short time, all gaps on the berm were filled by natural processes, and no evidence of the bulldozing could be seen. Fences and grass plantings were placed on the new dune face, and the pre-storm profile was reestablished. In most parts of town, all traces of the winter storm damage are now gone, and the dune has been rebuilt. As a result of beach grass planting and fencing, portions of the duneline are nearly four feet higher than in 1973, despite the storms of intervening years (fig. 15.2).

Public Education

The third major portion of Mantoloking's dune program is the education of residents and visitors regarding the need to protect the dunes. Many meetings and beach ecology walks have been conducted during the past four years to demonstrate how dunes grow, the role of beach plants in their growth and stabilization, the need to protect these plants, and techniques to assure their survival. Landowners have also been instructed in the care of their dunes during the dune inspection walks.

Another important education device is the town's local newsletter, *The Mantoloking Reporter*. Occasional articles deal with sources of plants, building of walkways, dune ecology, announcements of beach walks, and so forth. There is no doubt that an informative publication that appears regularly is a major stimulus to educating the public about local ecology and the dune system.

Conclusion

Mantoloking's dune program has been very successful, thanks to the efforts of a few dedicated individuals, followed by the combined efforts of the whole community.

The town's Dune Ordinance is strong and effective. It has been supported in the courts and has been shown to be in the public interest. The guidelines in the ordinance are continually revised as more information becomes available to the Council and the results of the dune program are observed. Regular physical measurements of the dune and beach system determine what changes are underway, and how effective the program has been. Lastly, an effective public educa-

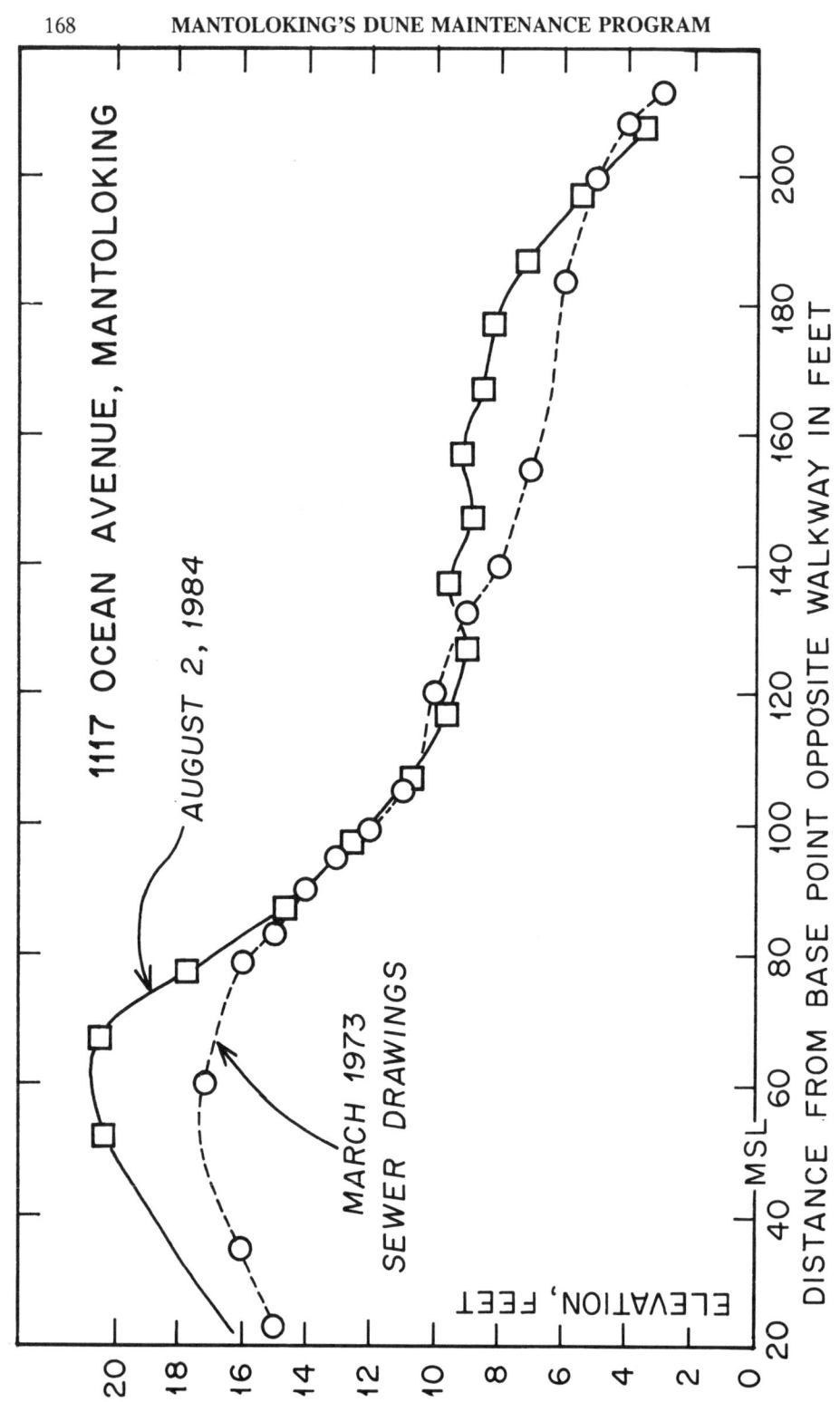

Figure 15.2. Dune and beach profiles showing growth of the dune between 1973 and 1984. Plotted by J. J. Serrell and redrawn by P. Robinson

tion effort through lectures, beach walks and outings, and an excellent newsletter, have all combined to make the whole community aware of the need to protect its dunes.

Residents of Mantoloking now know that the only thing standing between them and the ocean is that narrow strip of sand held together by beach grass. They realize that they must maintain the dune in order to reduce the probability that the ocean will inundate their town in the near future. What will happen over the long term, with a continually rising sea level and general retreat of the shoreline, remains to be seen. But by their concentrated community efforts, residents of Mantoloking have gained some time in which to enjoy their unique living situation.

References

Godfrey, P.J., and M.M. Godfrey, 1980. "Ecological Effects of Off-Road Vehicles in Cape Cod National Seashore." *Oceanus* 23: 56-67.

DUNE MANAGEMENT RECOMMENDATIONS FOR DEVELOPED COASTLINES

Paul A. Gares

Colgate University

Introduction

There has been a growing need for coastal researchers to provide reliable information about coastal systems that are modified by development and thus to contribute to the planning process for developed shorelines (Clark, 1978). One coastal system that deserves attention in this regard is the dune zone, which provides protection from flooding and buffers against shore erosion (Davies, 1957; Nordstrom and Psuty, 1980). Geomorphological investigations have provided information about the relationships between sediment movement and such independent variables as wind velocity and direction, sediment particle size, vegetation cover, and slope and elevation. These types of studies deal with the mechanics of aeolian sediment movement and provide an overall understanding of the processes that lead to dune formation. The problem is to determine the applicability of information derived in natural settings to developed shorelines, where human activities can alter the natural process-response relationships.

In an effort to contribute to the understanding of geomorphologic characteristics of developed coastlines, a comparative study was designed to examine factors that affect aeolian sediment transport in beach/dune systems along developed and undeveloped shorelines. The research objective of this study was to determine whether human actions associated with developed shorelines produce identifiable differences in the mechanics of sediment transport between the beach and the dune zone. This paper briefly reviews the results of the comparative study of developed and undeveloped shorelines and then summarizes specific management recommendations made to the New Jersey Department of Environmental Protection, based on results of the study.

Site Selection and Study Design

Study sites were selected in a variety of developed and undeveloped settings along Island Beach, New Jersey, a barrier spit that extends southward from Point Pleasant to Barnegat Inlet (fig. 16.1). This spit was selected for study because of the existence of adjacent developed and undeveloped areas. At the southern end of the spit is Island Beach State Park, which has remained almost completely undeveloped. The natural setting is interrupted only by recreational support structures. Two sites were selected for study in Island Beach State Park. Several municipalities are located on the barrier spit north of the park. They are characterized by a wide variety of land uses and housing densities. Five study sites were selected in the developed areas, representing different usage characteristics.

The study included both long- and short-term components. The first part of the study examined aerial photographs taken at regular intervals from 1952 to 1982 (Gares, 1983). Changes in position of the shoreline as represented by the high water line, the dune toe, and the dune crest were identified over the 30-year period. Differences in the amount of change at developed and undeveloped sites were attributed to activities associated with development and post-storm reconstruction. Special attention was given to the effects of the March 1962 Northeaster, which devastated the New Jersey shore.

The second part of the research comprised a field study in which sediment transport was monitored in different beach and dune environments through the use of vertical sand traps of the Leatherman (1978) design. A variety of sampling stations were selected at developed and undeveloped sites on the basis of topography, vegetation cover, and the presence of cultural features that could interfere with sediment transport. Sediment transport rates at developed and undeveloped sites were compared with the objective of identifying factors that might be responsible for any observed differences.

Summary of Findings

Long-term Changes

The use of earth-moving equipment, sand fencing, and vegetation planting are alternative means of restoring a damaged dune system. The effects of these two factors are demonstrated in the reconstruction activities conducted along the developed shoreline following the March 1962 Northeaster. A comparison of dune form six months after the storm to pre-storm conditions indicated that the dune toe had almost returned to its pre-storm position at site 3 (fig. 16.2). Studies have examined the growth of dunes as a result of vegetation planting or sand fence use (Woodhouse, 1978; Knutson, 1980), but neither of these approaches produced the

DUNE MANAGEMENT RECOMMENDATIONS

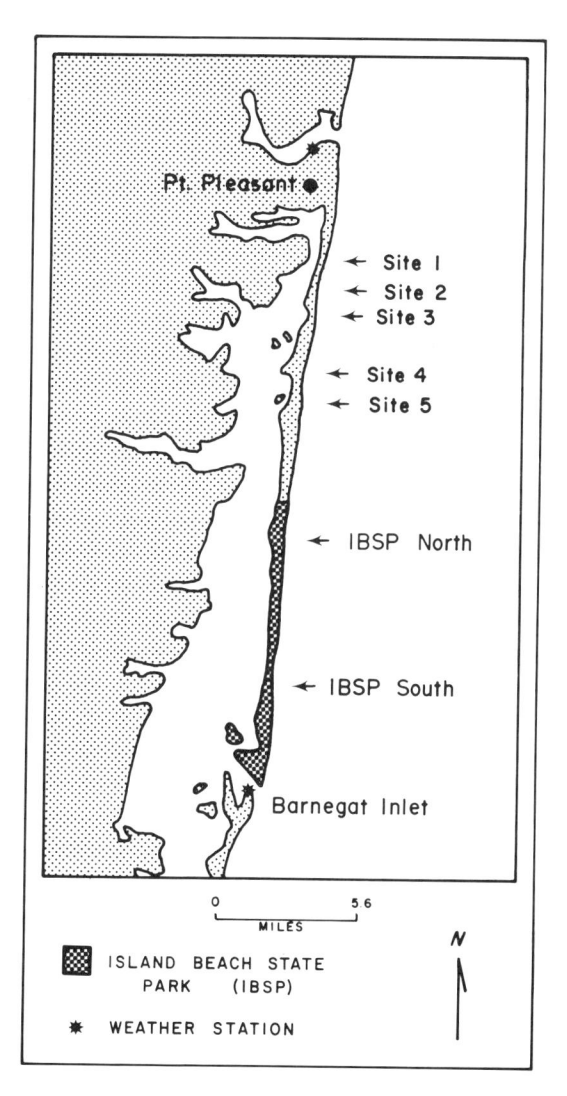

Figure 16.1. Island Beach spit, New Jersey, showing study sites

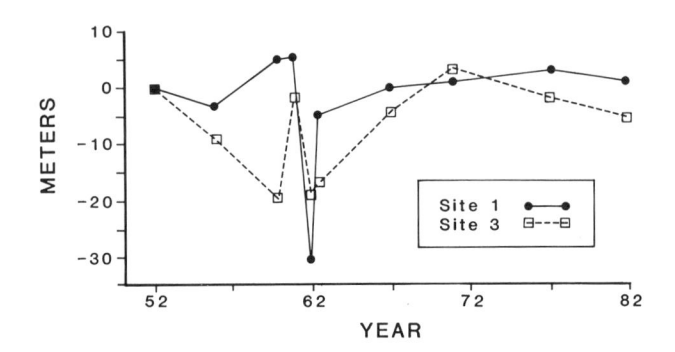

Figure 16.2. Cumulative changes in the position of the dune toe at sites 1 and 3, 1952-1982

amount of dune growth observed in this case. The logical conclusion is that the rapid regrowth of the dune following the 1962 storm could only have been produced by earth-moving equipment.

Five years after the storm, the dune toe had nearly returned to its pre-storm position at many of the study sites, including site 1 (fig. 16.2). This advance of the dune toe also appears to have resulted from human intervention, probably in the form of the installation of sand fences. Dune growth studies have shown that a substantial dune can be built in five to seven years through the use of sand fences (Woodhouse, 1978; Knutson, 1980). The same studies found that vegetation could produce a dune of comparable height but greater width in seven to ten years. The conclusion of these researchers, based on controlled experimental conditions, was that vegetation was just as effective as sand fences at building a dune over the long term, but that sand fences were more effective in the short term. It is doubtful that similar conditions existed along the developed portion of Island Beach spit in the years following the 1962 storm. These observations suggest that the seaward advance of the dune toe that occurred during the five-year period following the storm should be attributed to the use of sand fences, rather than to dune growth by natural processes alone. Indeed, dune toe advance at the undeveloped sites during the same five-year period was some eight times smaller than it was at the developed sites.

An examination of changes in dune area at developed sites points to some of the problems that can be associated with modifications in dune form caused by earth-moving equipment or sand fences. Reduction in dune area at all developed

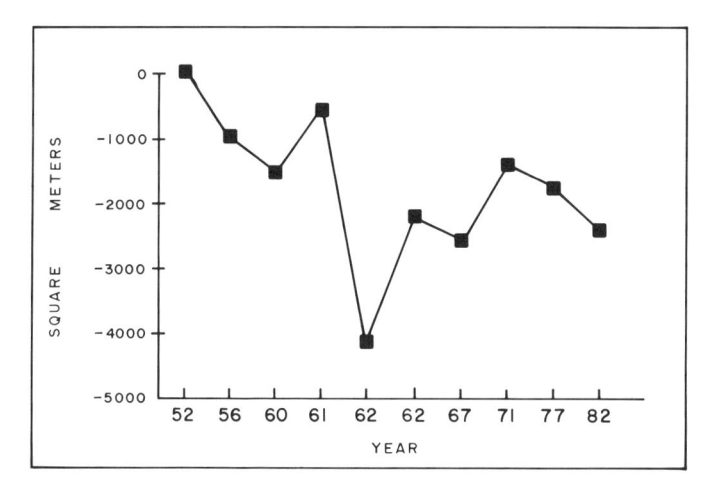

Figure 16.3. Average cumulative changes in dune area at developed sites for selected periods, 1952-1982.

sites was significant for the 30-year study period (fig. 16.3). The 1962 storm was responsible for a large amount of this reduction through erosion along the seaward face of the dune, but much of this erosion was repaired through dune building activities which extended the dune face seaward.

A logical expectation is that the dune area should have undergone little net change. In fact, the actual dune area grew back to only about one-half of its pre-storm size during the post-storm recovery period. Something must have happened along the landward margin of the dune to limit the recovery of the dune system. One conclusion is that the earth-moving equipment extended the size of the parking areas and driveways to the rear of beachfront properties at the expense of the existing dune. Also, construction of new homes in the dune zone after 1962 would have destroyed part of the dune system. Thus storm reconstruction activities and house building have important effects on the dune zone which are not evident from an examination of changes along the seaward face of the dune. These changes may appear inconsequential in an immediate time frame, but this reduction in the dune sediment budget reduces the amount of protection available over the longer term as the beach and dune system gradually narrows in response to sea level rise.

TABLE 16.1

SEDIMENT TRAPPED AT DIFFERENT BEACH SAMPLING STATIONS
November 2, 1981 – December 10, 1981
Values are given in cu. cm. per meter length of shoreline

Sampling Site	Volume Onshore	Volume Offshore
Undeveloped Sites		
IBSP 1	40229	6909
IBSP 2	56421	12875
IBSP 3	59378	65574
IBSP 4	49792	86224
IBSP 5	55860	30977
Developed Sites		
SITE 1B	22359	13742
SITE 1C	25418	4743
SITE 2A	32174	11524
SITE 3A	53947	6960
SITE 4A	18024	7775
SITE 5A	38269	12951

Short-term Changes

The data collected in the field study indicate that larger quantities of sand were transported in onshore and offshore directions at undeveloped sites than at developed sites (tables 16.1 and 16.2). These differences are primarily attributed to the use of sand fences in a variety of layouts and to variations in the vegetation density on the dune crest. Other factors such as topography, slope, or sediment particle size may also have affected the quantities of wind-blown sediment.

The amount of sand blown into the dune zone depends initially on the characteristics of the source area—in this case, the beach. Of the five developed sites, only sites 3 and 5 had onshore sediment transport rates on the beach that were comparable to those observed at undeveloped sites (table 16.1). Steeper beach slopes and coarser beach sediment may explain the lower sediment transport rates recorded at sites 1 and 2. The value for site 4A stands out from the beach values for developed sites because it is so small. This was a site where parallel rows of sand fences were erected on the beach perpendicular to the

TABLE 16.2

SEDIMENT MOVED THROUGHOUT THE SAME FENCE AT PAIRED LOCATIONS

November 2, 1981 - December 10, 1981

Values are given in cu. cm. per meter length of shoreline.

Sampling Site	Location	Volume Onshore	Volume Offshore
SITE 1B	Beach	22359	13742
SITE 1D	Dune Crest	20549	6882
SITE 1C	Beach	25418	4743
SITE 1E	Dune Crest	12593	6324
SITE 3A	Beach	53947	6960
SITE 3B	Dune Crest	587	484
SITE 4A	Beach	18024	7775
SITE 4B	Dune Crest	229	7775

northeast winds. The purpose for this layout was to trap sand moved onshore and to retain it as part of the beach profile. Sampling station 4A was located landward of the fences and reflects the degree of interference with aeolian sediment transport that results from the presence of the fences on the beach (table 16.1).

Although they were found on the beach at site 4, sand fences were most frequently used to control sand transport in the dune zone. The layout seen most often was a fence erected parallel to the trend of the shoreline at the toe of the dune. In general, sediment transport inland was only partially affected by a single thickness of fence. For example, dune crest sites 1D and 1E received 90% and 50%, respectively, of the sand moved inland from beach sites 1B and 1C (table 16.2).

When double lines of fence were used, sediment transport into the dune zone was virtually eliminated because the openings of one fence were filled by the slats of the other. This was the case at sites 3 and 4, where the fence formed an almost impenetrable barrier to onshore sand movement and allowed very little sand to be moved inland to the dune crest sites 3B and 4B (table 16.2).

The amount of vegetation cover on the dune crest was also found to be an important factor governing aeolian sediment transport in the dune zone. The dune crest at location 1D was bare, and this allowed a large proportion of the sand transported landward on the beach to be moved to the dune crest, despite the presence of the sand fence at the dune toe. By contrast, vegetation cover at station 1E was of moderate density, and this reduced the amount of sand transported onshore from the beach to the dune crest (table 16.2).

Vegetation was also important in determining whether the sand transported inland would be retained as part of the dune profile or whether it would be carried back in an offshore direction. A comparison of the two site 1 sampling stations demonstrates this difference. The amount of sediment moved offshore was comparable at dune crest locations 1D and 1E, but it was three times larger at beach site 1B than it was at beach site 1C (table 16.2). The difference in the amount transported offshore at the beach sites is attributed to the bare dune crest at site 1D. The presence of the sand fence at the dune toe did little on its own to stabilize the sand that was carried inland to site 1D. The moderate vegetation cover present at site 1E was much more effective at preventing offshore sediment transport. Thus, although the sand fence is effective at trapping sand moved inland, the presence of vegetation is more important in stabilizing this sediment.

The use of sand fences along the front of the dunes cannot be said to interfere with sand movement by wind, but they do prevent sand from moving very far inland. The use of fences thus results in the stabilization of dunes in a fixed location on the profile. Dune building practices may also extend the front of the dune onto the beach, narrowing the beach and placing the dune in a position of disequilibrium on the beach/dune profile, where it will be subject to increased erosion. The sand that is eroded from the front of the dune represents a loss in the sediment budget for the dune zone and, therefore, a loss of potential protection from storms. Given these observations, shoreline managers would do well to encourage the adoption of dune management practices that promote dune growth and maintenance along the shorefront.

Management Implications

Long-term shoreline management policy must recognize that sea level rise causes barrier shorelines to migrate gradually (Shepard, 1962; Kraft, 1971; Swift, 1975). Shoreline managers must decide whether to accommodate shoreline retreat or to resist it. If the selected approach is to accommodate shoreline retreat through policies which advocate dynamic shoreline management, the existence and growth of dunes along the shorefront is actively encouraged. The use of shore-parallel engineering structures is not acceptable under these circumstances because they may impede the landward migration of the dune zone. There are instances where protection structures are built landward of the dune in a way that allows the dune to migrate over them. Although this approach does not interfere with aeolian sediment transport in the short term, eventually the dune will move inland, leaving the structure on the beach. Under the management assumptions advocated here, the structure will have to be destroyed once it is uncovered. This stipulation makes it difficult to install permanent structures such as seawalls, because their size and cost would make it difficult to fulfill the management assumptions. It might be possible to build a bulkhead landward of the dune and then let the dune migrate over it. Under the dune, the bulkhead would represent additional protection should the dune be destroyed in a major storm. The bulkhead is a less massive structure than a seawall and represents less of a

DUNE MANAGEMENT RECOMMENDATIONS

TABLE 16.3

SUMMARY OF MANAGEMENT RECOMMENDATIONS
To Enhance Aeolian Sediment Transport and
Promote Dune Growth along Developed Shorelines

A. Enhance aeolian sediment transport
 1. Allow sand fences in the dune zone
 2. Plant vegetation for stability
 3. No fences on the beach
 4. No double fences
 5. Remove fences periodically

B. Construction practices
 1. No permanent protection structures
 2. Raise new houses and walkways on pilings
 3. Limit housing density at shorefront
 4. Establish setback zone
 5. No dune damage during house construction

C. Post-storm reconstruction
 1. Keep overwash sediment in dune zone
 2. No dune damage during reconstruction

---

commitment to maintaining a fixed shoreline position. Consequently, it would be easier to remove it once it becomes exposed on the beach.

Dune Management

Dune management recommendations presented in table 16.3 seek to promote a dynamic system by enhancing aeolian sediment exchange between beach and dune. This is achieved by regulating the configuration of shorefront structures and by controlling post-storm restoration of the beach/dune system. The results of this study indicate that sand fences have the most profound effect on the movement of sand by wind along developed shorelines. Sand fences are capable of producing the accumulation of sediment in the area immediately to the lee of the fence. A positive sediment budget for the dune zone may thus be created when fences are used along the seaward slope of the foredune, contributing to the growth of the dune. Therefore, the use of fences is to be encouraged.

Enhance Aeolian Sediment Transport

Although they do cause sediment accumulation, sand fences provide little long-term stability to the deposited sand (Woodhouse, 1978). In this study, newly deposited sediment was easily returned to the beach in areas along the dune crest which were bare or nearly bare. It is therefore important to plant vegetation on sediment accumulated through the use of sand fences in order to retain the sediment in the dune system and to produce a positive sediment budget for the dune zone. Once vegetation is planted to stabilize the sand, it will cause sediment to accumulate as effectively as sand fences (Woodhouse, 1978; Knutson, 1980).

Shorefront residents sometimes resort to using sand fences in ways that are not conducive to increasing the sediment budget of the dune zone. These practices should be discouraged. First, the use of fences or any other devices to accumulate or stabilize sand on the beach should be prohibited. The field data from this study indicate that fences erected on the berm significantly reduce the amount of sediment transported onshore. The sediment accumulated on the beach is easily removed during small storms because of its location on the profile. The sediment added to the beach budget provides no long-term protection to areas inland. This sediment would provide protection for a longer period of time if it were allowed to accumulate in the dune zone rather than on the beach.

A second practice often used by shorefront residents is to install new fences by attaching them to the old ones. The maintenance of the fence in good condition is necessary to ensure its continued ability to accumulate sediment. Attaching two fences together often compromises this goal when the alignment of the slats creates an impermeable wall. This configuration should be avoided.

Sand fences in the dune zone should be removed periodically to enhance uninterrupted sediment transport. This removal could take place once every four or five years and should last at least a full year. This action should only be taken if the dune crest has vegetation cover adequate to stabilize the sediment transported inland. In the absence of fences, onshore sediment movement should gradually cause the burial of the vegetation at the front of the dune, which will allow penetration of sediment further inland during subsequent storm events. The addition of sediment to the middle or rear of the dune crest will cause a gradual increase in sediment volume on the rear portion of the dune. This is an area that is rarely nourished when fences exist along the front of the dune, but it needs to receive sediment inputs if dune migration is to take place in accordance with the general management framework.

Construction Practices

The movement of sediment inland as a result of aeolian processes can also be facilitated by adopting regulations aimed at controlling development along the shorefront. Potential requirements might include maximum housing densities or some form of setback limit for the zone immediately adjacent to the beach. This would ensure the existence of space in which the dune can continue to form and eventually migrate. Another crucial requirement is to prevent the destruction of the dune during house construction. Although the dune could be pushed back into place after the house is finished, it would take time for the dune to grow and stabilize, leaving the area without storm protection for several years. This lack of protection is dangerous not only to the house immediately landward, but also to those houses that flank the property in question.

Finally, dune growth and migration can be enhanced by building new structures on pilings. The use of pilings to raise houses up over the surface of the dune is a requirement for National Flood Insurance coverage. It allows for space beneath the house to permit sediment to be transferred landward as the dune migrates (Nordstrom and McCluskey, 1985).

Post-storm Reconstruction

Post-storm reconstruction activities should also ensure that the integrity of the dune is not compromised. Sediment deposited in overwash fans during large storms must remain in the coastal barrier system. Under natural conditions, overwash fans deposited behind the dune become a source of sediment to be blown to the backslope of the dune, thus contributing to its future growth (Godfrey, 1976). In developed areas, overwash fans often interfere with human activities such as driving on a roadway. If overwash sediments must be removed, they should be returned to the dune zone rather than to the beach where they would be removed by waves. Natural dune nourishment with overwash sediment can be replicated by bulldozing overwash sediment to the backslope. Once there, it should be stabilized promptly with vegetation. This large-scale redistribution of sediment should only be undertaken where the overwash fans noticeably affect everyday human activities. It is appropriate for roadways, driveways, and parking areas, but not for other recreational areas surrounding homes, because the sediment is needed in those areas to nourish the backslope of the dune.

Another requirement for post-storm dune reconstruction involves limiting the loss of dune area. This study found that dune area was significantly reduced during the post-storm period by the expansion of driveways and parking areas, despite dune growth along the seaward face. Following a storm, the dune area in developed areas should show a loss that corresponds only to erosion along the seaward slope. With the passage of time, natural dune regrowth on the seaward side in conjunction with artificial building on the landward side should return the dune to its pre-storm size.

Conclusions

The construction of permanent facilities on the shoreline is not compatible with long-term coastal barrier dynamics. In high-hazard areas, structures are subject to flooding, damage by storm waves, or burial by migrating dunes. Even if buildings were raised on pilings to avoid these problems, the migrating barrier would eventually leave them stranded on the beach. In this location, houses would be in danger of destruction, and their presence on the beach would affect the recreational use of the beach.

The choice of time horizon is another issue for shoreline managers. One alternative is to overlook the problem of coastal barrier migration and let future generations deal with it when it comes to be perceived as a critical problem. Another alternative is to take steps to address the problem now in order to avoid a future crisis. Controlling both construction practices and post-storm reconstruction represents a logical way of reaching a compromise between the desires of some to develop the shoreline and the concerns of others regarding the future of these resources.

All persons involved with managing the shorefront have the same goal: to enhance the resource so that many people can enjoy it. The management of this zone should be seen as a cooperative effort on the part of individuals, municipal managers, and state managers alike. Dune management often relies on shorefront municipalities to take the initiative and institute dune protection ordinances. Although these local efforts are to be recommended, states should consider implementing incentive programs to encourage more communities to adopt dune management approaches such as those recommended here.

A large number of coastal barriers are heavily developed and management plans for these areas need to be made. Recommendations made here deal with three components of coastal barrier management: enhancing aeolian sediment transport, controlling current construction practices, and directing post-storm reconstruction. In each case, special consideration is given to facilitating the evolution of coastal dunes in such a way as to maximize the protection benefits they afford to inland areas. Activities such as using fences or vegetation in the dune zone are highly encouraged, whereas removal of sediment from the dune system using earth-moving equipment or preventing sand from being moved from the beach to the dune should be avoided. Longer-term practices which would enhance dune growth at the shorefront involve controlling house construction, both in terms of architectural style and spatial location. Dunes require space in which to function. Current building practices reduce that space.

Acknowledgments

This project was funded by the New Jersey Department of Environmental Protection, Division of Coastal Resources, through the Center for Coastal and Environmental Studies at Rutgers—The State University of New Jersey. The author would like to thank Susan Halsey of NJDEP and Norbert Psuty and Karl Nordstrom of CCES for their reviews of the research and for their suggestions for improvement.

References

Clark, M.J. 1978. "Geomorphology in Coastal Zone Environmental Management." *Geography* 63(4): 273-282.
Davies, J.H. 1957. *Dune Formation and Stabilization by Vegetation and Plantings.* CERC Technical Memorandum #101. Ft. Belvoir, Va.: U.S. Army Corps of Engineers.
Gares, P.A. 1983. "Beach/Dune Changes on Natural and Developed Coasts." In *Proceedings of Coastal Zone '83.* New York: American Society of Civil Engineers, pp. 1178-1191.
Godfrey, P.J. 1976. "Barrier Beaches of the East Coast." *Oceanus* 19(5): 27-40.
Knutson, P.L. 1980. *Experimental Dune Restoration and Stabilization, Nauset Beach, Cape Cod, Massachusetts.* CERC Technical Paper #80-5. Ft. Belvoir, Va.: U.S. Army Corps of Engineers.
Kraft, J.C. 1971. "Sedimentary Facies Patterns and Geologic History of a Holocene Transgression." *Geological Society of America Bulletin* 82: 2131-2158.
Leatherman, S.P. 1978. "A New Aeolian Sand Trap Design." *Sedimentology* 25: 303-306.
Nordstrom, K.F., and N.P. Psuty. 1980. "Dune District Management: A Framework for Shorefront Protection and Land Use Control." *Coastal Zone Management Journal* 7(1): 1-23.
Nordstrom, K.F., and J.M. McCluskey. 1985. "The Effects of Houses and Sand Fences on the Eolian Sediment Budget at Fire Island, New York." *Journal of Coastal Research* 1: 39-46.
Shepard, F.P. 1962. "Gulf Coast Barriers." In *Recent Sediments Northeast Gulf of Mexico.* Tulsa: American Association of Petroleum Geologists.
Swift, D.J.P. 1975. "Barrier Island Genesis: Evidence from the Central Atlantic Shelf, Eastern U.S.A." *Sedimentary Geology* 14: 1-43.
Woodhouse, W.W., Jr. 1978. *Dune Building and Stabilization with Vegetation.* CERC Special Report #3. Ft. Belvoir, Va.: U.S. Army Corps of Engineers.

INDIVIDUAL ATTITUDES TOWARD COASTAL EROSION POLICIES: CAROLINA BEACH, NORTH CAROLINA

Owen J. Furuseth and Sallie M. Ives

University of North Carolina at Charlotte

Introduction

To most Americans, coastal erosion is not a serious natural hazard. Erosion poses no direct threat to human life and it is a low-frequency, continuous hazard which tends to affect areas used primarily for recreation (Mitchell, 1974). Nevertheless, erosion is responsible for losses in real estate property values and for disruption of coastal ecosystems. Estimates of the annual national losses caused by erosion vary from $185 to $235 million, with significant erosion affecting roughly one-quarter of the U.S. shoreline (U.S. Army Corps of Engineers, 1971). Erosion also plays havoc with legal property rights in coastal real estate.

Consequently, it is not surprising that where serious shoreline recession is juxtaposed with densely settled coastline or areas experiencing strong, urban-related development pressure, local economic interests often view coastal erosion as a near-catastrophic natural hazard. This situation usually results in lobbying for government-funded erosion control and shoreline protection assistance. While the federal government, through the U.S. Army Corps of Engineers, is able to provide varying degrees of assistance, the requests for aid far exceed funding levels. More importantly, long-term, cost-effective erosion control strategies are not, and may never be, available. The best solutions for many communities and individual property owners are extremely costly and only temporarily effective physical adjustments. Usually, communities opt for the erosion control strategy that costs them least. This may not, however, be the lowest cost or most "cost effective" solution from a general societal viewpoint.

Over the past 50 years, much research has been focused on coastal erosion, with most of the emphasis on structural remedies for the problem. Despite urgings that behavioral and attitudinal research offers "promising prospects for research payoffs" (Sorensen and Mitchell, 1975), few investigators have explored community attitudes and support for various erosion adjustments.

In light of the failure of existing technologies to control coastal erosion, it seems reasonable that we should try to understand individual responses to the erosion hazard and the associated adjustments. How do coastal residents respond to the various policy adjustments for minimizing the threat of erosion? In virtually every community affected by coastal erosion, there are interests seeking federal and state financial assistance. Is coastal erosion hazard in fact a high-priority concern to the community as a whole? In an era of increasing budgetary constraints and environmental sensitivity, better knowledge of local perception of erosion could be valuable in assessing the need for various erosion control strategies and targeting where limited technical and fiscal resources would generate the greatest benefits.

The Study Area: Carolina Beach

The research reported in this paper represents the second phase of an empirical analysis of community and individual responses to coastal erosion hazard in Carolina Beach, North Carolina. The study area is a small, recreational and maritime-oriented community, typical of many coastal barrier communities along the Atlantic and Gulf coasts (fig. 17.1). Until quite recently, Carolina Beach remained a traditional coastal resort, characterized by small, locally owned motels and privately owned beach cottages. However, during the past five years and especially in the past 18 months, the pace of new condominium development has quickened and Carolina Beach has become a commercialized resort area.

Dating back to the late 1800s, Carolina Beach has suffered from a long history of coastal storms and erosion hazard. Since 1952, the town has suffered severe erosion problems. The increase in beach erosion followed the construction of a tidal inlet some 7,500 feet north of the town limits by the U.S. Army Corps of Engineers. Ironically, the dredging of the inlet was initiated and completed at the request of Carolina Beach recreational interests and the local government. Following the completion of this project, the shoreline erosion rate increased from an average of 0.6 to 2.8 feet per year. Subsequently, the Corps of Engineers has engaged in an ongoing, periodic effort to protect and nourish the foreshore. A major beach nourishment project was completed in July 1982, involving 3.66 million cubic yards of sand at a cost of $8.8 million. Unfortunately, this project was followed by an unusually severe winter season. As a result, large quantities of the newly created foreshore were lost to the sea as "northeaster" storms hit the area from December through March.

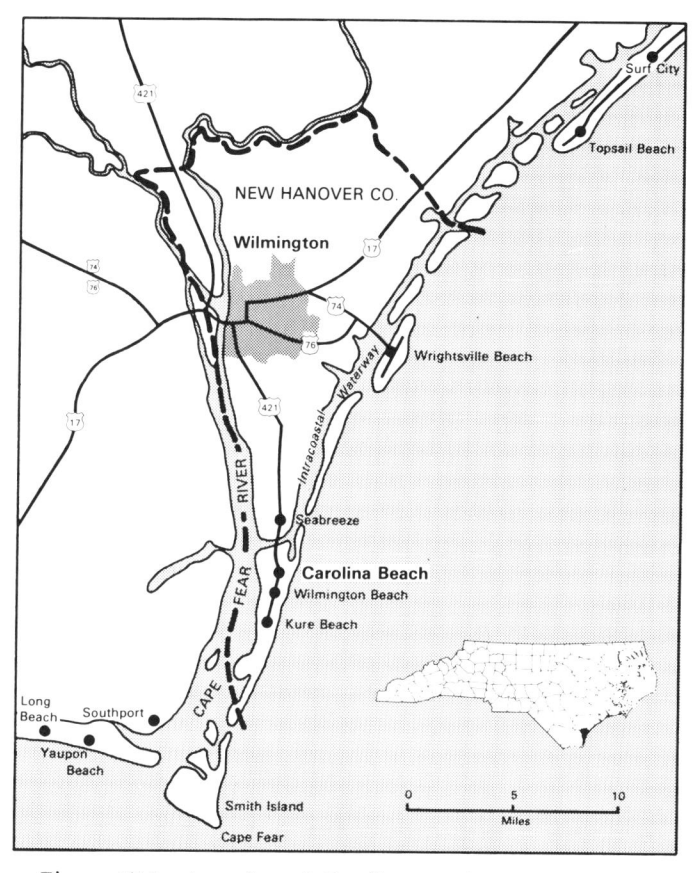

INDIVIDUAL ATTITUDES TOWARDS COASTAL POLICIES   187

Figure 17.1. Location of Carolina Beach, North Carolina

Community Attitudes and Perception of Erosion Hazard

In March 1983, the authors and two graduate assistants conducted personal interviews designed to measure community attitudes, perception of erosion hazard, and knowledge of current erosion adjustments. Ninety-one interviews were completed, representing an approximate 5% sample of the 2,000 year-round residents. The first phase of the analysis concerned three main issues: (1) How does the community perceive the coastal erosion hazard? (2) What are community attitudes toward the adjustments to beach erosion? (3) What is the perceived role of government in mitigating erosion risk? (Furuseth and Ives, 1984)

Among the most significant findings was that Carolina Beach residents are very knowledgeable about the causes of coastal erosion, but somewhat deterministic in their attitudes toward the hazard. Moreover, we found a high level of community awareness and sensitivity to erosion hazard, far exceeding the level of cognition found in previous research by Mitchell (1974) and Rowntree (1974). Also somewhat surprising were the findings regarding preferred adjustments to erosion. The respondents strongly favored nonstructural approaches, including local land-use planning and building regulations to control development in erosion-risk areas, with slightly less support for financial assistance from state and federal governments to repair erosion-caused damage. The least popular form of adjustment was the use of erosion control structures, including groins, jetties, and seawalls.

It appears that community attitudes in Carolina Beach do not fit the popular image which suggests that coastal residents prefer erosion control structures paid for by federal or state governments. Rather, most Carolina Beach residents are aware of the erosion hazard and tend to view it as a continuous, natural process. They recognize that the risk of erosion is increased by human action, particularly actions involving encroachment onto the foreshore. Finally, they accept the futility of trying to control erosion, and accept land-use management designed to accommodate beach recession.

Individual Attitudes Toward Erosion Policies

Having examined community-wide perception and attitudes toward coastal erosion and erosion mitigation policies, the second phase of our analysis focused on the variation in individual attitudes toward erosion management options. Are there significant differences across the community in attitudes and awareness toward erosion mitigation policies? If so, what factors account for the differences? These questions constitute the substantive issues analyzed here.

Previous research has shown that individual responses to risks of natural hazards and the choice of adjustments are often related to hazard experience, economic status, tenure characteristics, and personality factors (White, 1974). Our analysis considered the effect of 11 locational, tenure, and socioeconomic characteristics on the responses to erosion mitigation policies. Our purpose was to test the hypothesis that individual attitudes are influenced by the social and residential characteristics of the respondents.

Methods and Findings

While there are a number of possible individual adjustments for those wishing to mitigate coastal erosion hazard, for the purpose of this analysis, respondents' attitudes toward three classes of erosion mitigation policy were examined. These include engineering and physical strategies designed to control or reduce erosion, government-sponsored relief and rehabilitation assistance, and police power approaches to control erosion risk. The selection of these policy options was based on the widespread use of these policies by most coastal communities.

In order to measure attitudes toward these adjustments, attitudinal scales were developed for each policy option. The scales were constructed by combining three to five Likert-type questions measuring individual responses to a specific policy option (table 17.1). The questions were taken from the Carolina Beach survey questionnaire. All of the questions had a response format of "strongly agree," "agree," "neither agree or disagree," "disagree," and "strongly disagree" and were coded as 5, 4, 3, 2, and 1, respectively. Where wording required, answers were recoded. The total number of points on all questions was tallied to derive the final scores. Subsequently, a high score indicated a positive attitude toward a specific policy and a low score, an unfavorable attitude.

Prior to the selection of the final questions used in each scale, the questions were examined for internal consistency. Only those questions with a high degree of intercorrelation (r values exceeding .70) were included in the final scale. Those questions not meeting this criterion were considered as measuring different concerns and were therefore not included in the analysis.

Following the construction of the attitudinal scales, the scores for each policy option were tallied and their distributions examined. They were then recoded to fit an ordinal scale and to create a normal data distribution. Subsequently, a contingency table was constructed so that the scores could be compared with the 11 predictor variables. The degree of association between individual variables and scores is expressed using tau (t) and theta (0) coefficients. The tau coefficient is commonly used to measure the association among ordinal scale variables, while theta is used to describe the association between ordinal and nominal scale data. As with other coefficients of association, t and 0 values range from 0 to 1. The sign of both coefficients can be interpreted as indicating the direction of the relationship.

Finally, in order to test the research hypothesis, the tau (t) and the Mann-Whitney U-tests (U) of significance were calculated for contingency table. The tau test of significance was applied to the ordinal scale cross tabulations, with the U

TABLE 17.1

QUESTIONS USED TO DEVELOP THE EROSION POLICY ATTITUDES SCALE

Policy Approach	Questions
Government Assistance to Reduce Erosion Risks and Compensate Affected Property Owners	The federal government should provide money to repair beach erosion damage.
	The state should provide money to repair beach erosion damage.
	Beach erosion is the problem of the individual property-owner and taxpayers should not have to pay for solutions to the problem.*
	The federal government should provide inexpensive insurance for anyone who would be affected by beach erosion.
Police Power Strategies to Control or Reduce Erosion Risks	The federal government should control the use of land along beaches.
	Local governments should use zoning and other regulations so that buildings are not put in areas that erode.
	Stronger laws are needed to prevent the destruction of dune vegetation.
Engineering and Physical Improvements to Control or Reduce Erosion	Replacing sand (replenishment) on an eroded beach is only a temporary measure.*
	Seawalls are the best long-term option for saving a beach.
	The local government should build groins and jetties to reduce beach erosion here.
	Seawalls destroy rather than save a beach.*
	Most of the beach erosion here would be stopped if sand fences and other stabilization structures were put in the correct places.

*In accordance with the research methods this question was reverse coded.

test used for the ordinal and nominal scale cross tabulations. Both tests are equivalent in function and thus permit comparable conclusions.

The results of the analysis, presented in tables 17.2-17.4, show varying support for the research hypothesis. Clearly, individual attitudes toward most erosion

TABLE 17.2

SUMMARY OF RELATIONSHIPS BETWEEN VARIABLES AND
GOVERNMENT ASSISTANCE TO REDUCE EROSION RISKS OR
COMPENSATE AFFECTED PROPERTY OWNERS

Independent Variables	Test of Significance Results		Degree of Association	
	t	U	t	θ
location of residence				-.028
length of residency			-.005	
tenure				-.091
value of residence			.072	
type of residence				-.021
age	*		-.467	
gender				.071
marital status				-.078
income			-.033	
educational attainment			-.029	
political orientation			-.030	

SOURCE: Calculated by the authors.
*statistically significant beyond the .95 level.
**statistically significant beyond the .90 level.

policies are not predictable based upon socioeconomic, locational, or residential characteristics. Nevertheless, there are substantial differences in the effect of these variables between different erosion policies, and these differences warrant consideration.

Government assistance to reduce erosion risk and compensate affected property owners displayed the weakest association (table 17.2). Among the predictor variables, only age is statistically significant. The tau coefficient (-.467) indicates that approximately 47% of the variation in attitude toward this policy option is accounted for by a respondent's age, with younger respondents favoring this approach more and older residents favoring it less.

One explanation for the ineffectiveness of most variables in differentiating attitudes toward this policy approach may be that relief and rehabilitative assistance is a popular and long-standing public policy. This type of policy "socializes" environmental risk so that the cost of assistance is not born by the affected individual property owners, nor the community alone (Rossi et al., 1982); rather, all taxpayers assume a small share of the cost. Moreover, the community as a

TABLE 17.3

SUMMARY OF RELATIONSHIPS BETWEEN VARIABLES AND
POLICE POWER APPROACHES TO CONTROL OR REDUCE EROSION RISKS

Independent Variable	Test of Significance Results		Degree of Association	
	t	U	t	θ
location of residence				-.090
length of residency	*		-.292	
tenure				-.031
value of residence			-.044	
type of residence		*		.243
age			.028	
gender				.055
marital status				-.063
income			.001	
educational attainment			.100	
political orientation			.001	

SOURCE: Calculated by the authors.

*statistically significant beyond the .95 level.

**statistically significant beyond the .90 level.

whole benefits from state or federal assistance. With this in mind, it is not unexpected that this policy has widespread support from the community.

Similar results are evident in the attitudes toward police power controls (table 17.3). Among the predictor variables, only length of residence and type of residence are statistically significant. The operational direction of these variables indicates that newer residents to Carolina Beach and those residents not living in detached, single-family houses are most supportive of police power regulations. Excepting these two residential factors, none of the other variables proves to be a valuable predictor of attitudes.

It is particularly noteworthy that the political orientation and income variables showed especially weak results. In general, most of the literature indicates that these two variables are important and consistent predictors of attitudes toward police power policies, with political conservatism and higher income marked by strong opposition to land-use controls (Reilly, 1973). In Carolina Beach, however, political orientation and income make no significant difference in how a respondent evaluated erosion-related police power actions.

Again, the absence of any significant relationships among a large number of variables suggests that within Carolina Beach there is only a minimal amount of

TABLE 17.4

SUMMARY OF RELATIONSHIPS BETWEEN VARIABLES AND
ENGINEERING AND PHYSICAL IMPROVEMENTS TO
CONTROL OR REDUCE EROSION

Independent Variable	Test of Significance Results		Degree of Association	
	t	U	t	θ
location of residence		**		-.153
length of residency			-.037	
tenure		**		-.142
value of residence	*		.224	
type of residence				-.086
age	**		-.113	
gender		**		.145
marital status				-.086
income			.011	
educational attainment	*		-.204	
political orientation			-.021	

SOURCE: Calculated by the authors.
*statistically significant beyond the .05 level.
**statistically significant beyond the .10 level.

variation in attitudes related to the predictor variables. Although attitudes toward police-power approaches are not uniform across the community, they cannot be explained by these factors.

In contrast to the previous policy approaches, the analysis of engineering and physical controls shows a substantial increase in variable performance (table 17.4). Over half of the variables are statistically significant predictors. Value of residence and educational attainment are the strongest, with the greatest support for engineering and physical policies among residents with the most expensive homes and the lowest educational attainment. Also statistically significant, but with weaker reliability, are location of residence, tenure, age, and gender. Unlike the previous analyses, these results would seem to provide moderate support for the research hypothesis. The data suggest that, in the case of engineering and other physical approaches to coastal erosion, socioeconomic and residential characteristics influence personal appraisal of the policy. Thus, in Carolina Beach, the supporters of beach nourishment or jetty construction are most likely to be older, less educated, and female beachfront homeowners. In contrast, residents who are less supportive of this strategy tend to be younger, better-educated persons living in apartments, duplexes, or lower-value, detached residences not along the beachfront.

One interpretation of these findings may be that for those respondents who feel at greatest risk from erosion, this policy approach provides a sense of protection, while less threatened respondents tend to view these approaches as less necessary and only temporary solutions. This interpretation is partially supported by earlier investigations of Mitchell (1974) and Rowntree (1974), who found that shorefront property owners and managers tend to be deterministic in their perception of erosion hazard and view engineering approaches as vital to their well-being. Conversely, Rowntree also found that less threatened, non-shorefront respondents are far less supportive of erosion control measures. This is not surprising, since they are not directly affected by either the hazard or the mitigation action.

Discussion and Conclusion

Analysis of attitudes among Carolina Beach residents with respect to alternative erosion mitigation policies, while not completely supporting the research hypothesis, does provide insight and raises several substantive issues. The first concerns the absence of consistent performance by any single predictor variable. None of the 11 residential or socioeconomic factors were statistically significant for all three policy approaches, and only one variable, age, was significant in two cases. For Carolina Beach, at least, this indicates that different characteristics are important in separating and understanding attitudes for different policies. There is no single characteristic or set of characteristics which affects the perception of all erosion policies. One inference that may be suggested by this weak relationship is that individual responses toward different erosion mitigation policies are affected by different dimensions of concern and interest. That is to say, attitudes toward one policy may be derived from one set of perceptions and concerns, while attitudes toward a second policy are affected by completely different interests.

Finally, the data indicated that, in Carolina Beach, response to alternative erosion mitigation strategies varies with the type of policy. Community attitudes toward government assistance and police power policies are rather consistent, with no noteworthy attitudinal differentiation based on community characteristics. This is not the case, however, for engineering or physical approaches to erosion, which generate considerable variance in response.

On the basis of these differences, it may be suggested that there is a measure of consensus within the community with respect to government assistance or police power actions, but much less consensus with respect to engineering or physical actions. Subsequently, the pursuit of the latter policy could have far greater associated social impacts, particularly in coastal areas already stressed by rapid growth. The potential for introducing conflict into a community by pursuing alternative erosion control policies is a topic that warrants serious consideration by researchers and decision makers.

Our research is limited to a single coastal community; therefore, the findings lack direct transferability to other coastal towns and settings. Nevertheless, these findings provide a starting point from which to address the many questions concerning community attitudes and perception of coastal erosion policies. Without an understanding of public response to these actions, it is impossible to develop technically and socially acceptable adjustments. Failure to expand on this research framework and to consider the social and behavioral dimensions affecting erosion policy development will result in a continuation of our currently muddled efforts.

Acknowledgments

The authors wish to acknowledge partial financial support from the Natural Hazards Research and Applications Information Center (Boulder, Colorado) through a National Science Foundation grant and also wish to thank Anne Garren and Suzanna Schwartz for their help in data collection.

References

Furuseth, O.J., and S.M. Ives. 1984. "Community Response to Coastal Erosion Hazard." *Southeastern Geographer* 24, no. 1 (May), pp. 42-57.
Mitchell, J.K. 1974. *Community Response to Coastal Erosion*. Research Paper no. 156. Chicago: Department of Geography Research Paper no. 156, University of Chicago.
Reilly, W.K. 1973. *The Use of Land: A Citizen's Policy Guide to Urban Growth*. New York: Thomas Y. Crowell.
Rossi, P.H., J.D. Wright, and E. Weber-Burdin. 1982. *Natural Hazards and Public choice, the State and Local Policies of Hazard Mitigation*. New York: Academic Press.
Rowntree, R.A. 1974. "Coastal Erosion: The Meaning of a Natural Hazard in the Cultural and Ecological Context." In *Natural Hazards: Local, National, Global*. Ed. Gilbert White. New York: Oxford University Press.
Sorensen, J.H., and J.K. Mitchell. 1975. *Coastal Erosion Hazard in the United States: A Research Assessment*. Boulder: Institute of Behavioral Science, Program of Technology, University of Colorado.
U.S. Army Corps of Engineers. 1971. *Report on the National Shoreline Study*.

Washington: Crops of Engineers.

White, G.F. 1974. "Natural Hazards Research: Concepts, Methods, and Policy Implications." In *Natural Hazards: Local, National, Global.* Ed. G.F. White. New York: Oxford University Press, pp. 3-16.

HAZARD MANAGEMENT

DECIDING WHETHER TO EVACUATE A BEACH COMMUNITY DURING A HURRICANE THREAT

Jay Baker

Florida State University and
Hazards Management Group, Inc.
Tallahassee, Florida

The Evacuation Decision Dilemma

If coastal residents and vacationers are to be safe from hurricanes, they must escape the storm surge that accompanies such storms. Storm surge is an increase in sea level caused by storm forces, often resulting in water levels several feet higher than normal. The traditional means of escaping is by evacuation, that is, going inland to a safe elevation some distance from the shore. This action is particularly important in coastal barrier areas that receive the brunt of the surge, scour, and wave battering.

The length of time necessary to safely evacuate a coastal population increases as the population expands, assuming that transportation capabilities remain constant. Quantitative estimates of those times were largely guesswork until fairly recently. In 1978, Urbanik calculated the time necessary to evacuate Galveston, Texas and, in 1976, evacuation time estimates were calculated for Sanibel, Florida as part of the city's growth management plan. In Lee County, Florida in 1979, for the first time, evacuation times were calculated as part of a comprehensive hurricane evacuation plan, which both improved the plan and gave more thoroughly conceived estimates of the times (Griffith, 1979).

Since the Lee County study, comprehensive, hurricane evacuation studies have been completed in several areas, usually involving multi-county regions. In some locations, estimates of evacuation times exceed 24 hours, even for Category 3 hurricanes (Tampa Bay Regional Planning Council, 1984; Ruch, 1983). The estimates include mobilization time after an order has been issued, travel time, and pre-landfall hazards time (usually the time between onset of gale conditions and the arrival of the eye's closest point of approach). These numbers indicate how long before arrival of the hurricane's eye the evacuation must begin to reasonably ensure the safe evacuation of everyone at risk.

Clearly this is valuable information for emergency preparedness professionals and public officials responsible for public safety. Several problems remain, however:

(1) Twenty-four hours before landfall at its eventual destination, no one knows where the storm is going to hit. The average forecast error for the 24-hour position of a storm is about 100 miles, meaning that the warning area for the hurricane (the broad area within which the storm is almost sure to hit) will be about 300 miles wide. Only one-fourth or less of that area will be severely affected, however. Thus, if a barrier beach area is evacuated when the storm is still 24 hours away, the evacuation will prove to have been unnecessary in perhaps three out of four cases.

(2) In forecasting the position of a storm, there is at least as much error in predicting how fast it will move along a track as there is in forecasting the track itself. This impedes estimation of when the storm is 24 (or 36 or 18) hours away, obviously complicating the use of evacuation time estimates in deciding when and whether to order the evacuation of a community.

(3) Severity of the storm is another uncertainty that will affect not only whether an area has to be evacuated, but also how early the evacuation must begin (as severe storms necessitate the evacuation of larger areas and more people). One concern is the spatial extent of winds of a particular velocity, as well as peak winds and minimum pressure levels.

The public depends upon public officials (federal, state, county, local) to decide whether and when evacuation is necessary (Baker, 1983a). Relatively few people will leave until advised or ordered by public officials. Thus those officials and the professionals upon whom they depend have not only a difficult task but a highly responsible one as well.

Some communities are beginning to consider vertical evacuation or refuge as an alternative to conventional evacuation. "Vertical evacuation" means that the population at risk seeks shelter in multi-story buildings above the storm surge level rather than trying to evacuate inland. This can be used to postpone a decision about evacuation until officials are quite certain that the storm is going to hit. "Vertical refuge" is similar to vertical evacuation, but uses multi-story buildings only as a last resort. This option gives officials a perceived margin of safety if the decision is made to hold off an early evacuation longer than might normally be judicious. The pros and cons of vertical evacuation and vertical refuge are considered in more detail elsewhere (Baker, 1983b).

Information Available for Decision Makers

Evacuation Studies

As part of the hurricane evacuation studies mentioned earlier, scores and sometimes hundreds of hurricanes are simulated with SLOSH or SPLASH computer models for their effect on the study area. The simulated storms vary with respect to severity, landfall location or closest point of approach, angle of approach, forward speed, and whether they are landfalling, paralleling, or exiting with respect to land areas. A threatening hurricane can usually be matched with one of the simulated storms to provide estimates of the real storm's effects and implications for evacuation of locations in the study area. The chances of matching them correctly is likely to be no better than our ability to forecast the threatening storm's behavior.

National Hurricane Center Alert

When a hurricane begins to threaten a coastal area, the National Hurricane Center (NHC—a unit of the National Weather Service) issues a hurricane watch for a broad reach of coastline—probably covering several hundred miles. The storm may hit somewhere in that area within the next 36 or 48 hours, and the watch area can shift spatially as the storm moves.

When a storm is expected to make landfall within the next 24 hours, a hurricane warning is issued. The warning area will typically be about 300 miles wide and might shift as forecasts of storm movement change. In recent years, the average period of warnings has been 19 hours rather than 24. Watches and warnings have been in use for many years, and most emergency management professionals are familiar with them.

Forecast Storm Position

The watch and warning areas are very wide; it would be useful to know where within these areas the storm is most likely to strike. The watches and warnings give only the vaguest notion of landfall timing. The NHC forecasts where the storm center is most likely to be 12, 24, 48, and up to 72 hours after its current position. However, without proper consideration of its reliability (or lack thereof), the forecast position can be very misleading and dangerous.

The longer-range the forecast, the greater its error is likely to be. The average position error can be roughly estimated by multiplying the number of hours in the forecast by five or six. Thus the average 24-hour forecast is usually off by about

125 miles, the 48-hour forecast by about 250 miles, and so forth. Figures 18.1-5 show various forecasts for hurricane Eloise in 1975 (Carter, 1984). In each figure there are four ellipses containing a center point and other points not inside an ellipse. The points standing alone are past positions of Eloise, and the point immediately to the right or below the smallest ellipse is the "current" position of the storm. The point inside the smallest ellipse is where the storm is predicted to be 12 hours after its current position; the other points represent the 24-, 48-, and 72-hour forecast positions of the storm. The radii of the ellipses represent the average forecast error for the respective time period. The actual 12-hour position of the storm will be inside the ellipse for that time period about 65-70% of the time.

Figure 18.1 shows the storm's position at 2 p.m. EDT on September 17, 1975, and where it was predicted to be 12, 24, 48, and 72 hours later. Figure 18.2 shows its position and forecasts at 8 a.m. on September 20. All the intermediate forecasts plotted the storm taking a northerly turn farther east than this one, but Eloise persisted in its westerly track. As figure 18.3 indicates, forecasters finally accepted the westerly path at 2 p.m. and rerouted the storm into Mexico rather than the northern Gulf Coast, only to have the storm then turn to the north and force a return to the earlier forecast pattern (fig. 18.4). The ultimate track of the storm is shown in figure 18.5. Landfall occurred to the west of Panama City, Florida around 7 a.m. on September 23. Residents were not advised or ordered to evacuate by local officials until after midnight, just seven hours before landfall.

The forecast track of Hurricane Frederic in 1979 was much closer to the actual path, but it illustrates the error that can occur in forecasting the forward speed of a storm. Figure 18.6 gives Frederic's position at 5 p.m. EDT on September 11, 1979; it was forecast to make landfall in just over 48 hours. NHC advisories normally come out every six hours, but by 9:30 p.m., a new advisory was issued that Frederic would make landfall in just over 24 hours (fig. 18.7). At 5 p.m., emergency management professionals had thought they had two full days to get ready if Frederic came their way; yet in the next four and a half hours, they had "lost" a full day (Carter, 1983).

Severity Forecasts

As difficult as position forecasts are, forecasts of intensity are even more difficult. Figure 18.8 shows the peak wind velocities of Hurricane Alicia over a three-day period before landfall occurred at Galveston, Texas. From 5 p.m. on August 16, 1983, through 8 a.m. on August 17, Alicia's peak winds maintained a constant 80 mph. Toward the end of this period, NHC public advisories said only that intensification was "possible." At 11 a.m., peak winds began to climb, and by 5 p.m. they were 110 mph and advisories were calling Alicia a dangerous storm. A decision by Galveston officials had been made earlier not to order evacuation of the island because of the storm's relatively weak intensity; by the time the storm appeared to pose a threat (particularly if intensification had increased further), complete evacuation would have taken too long to complete before landfall.

THE EVACUATION DECISION

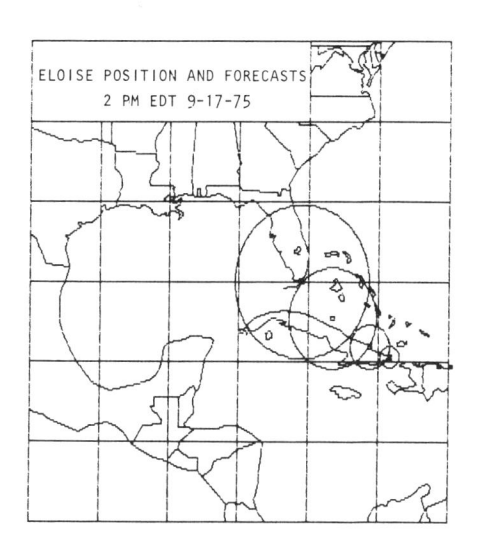

Figure 18.1

Figure 18.2

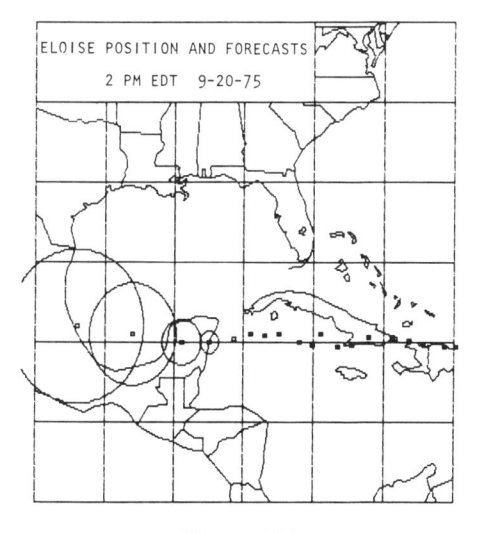

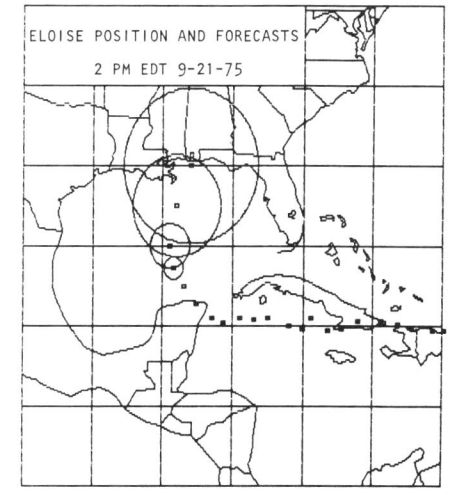

Figure 18.3

Figure 18.4

204   THE EVACUATION DECISION

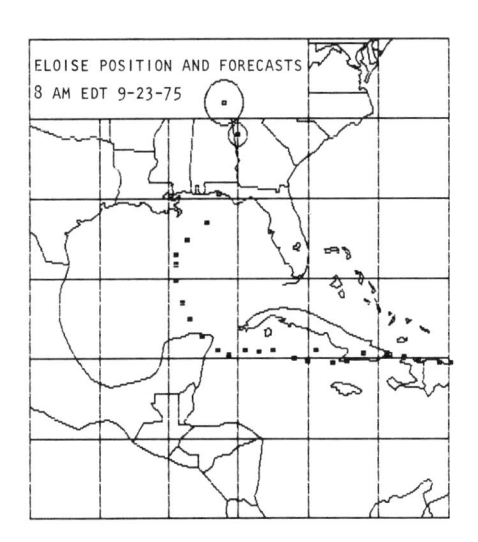

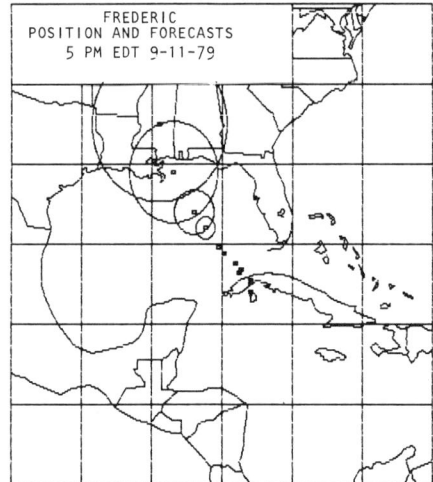

Figure 18.5

Figure 18.6

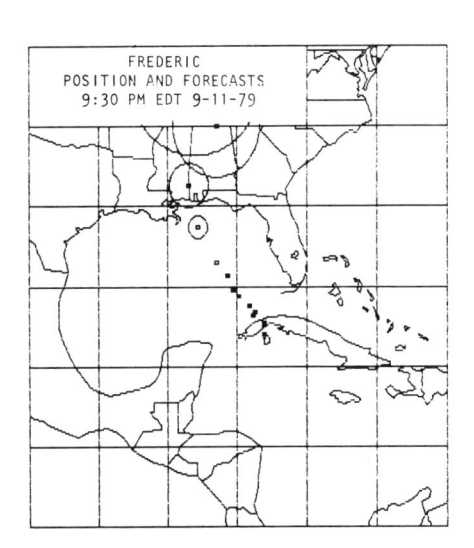

Figure 18.7

THE EVACUATION DECISION

ALICIA'S PEAK WINDS

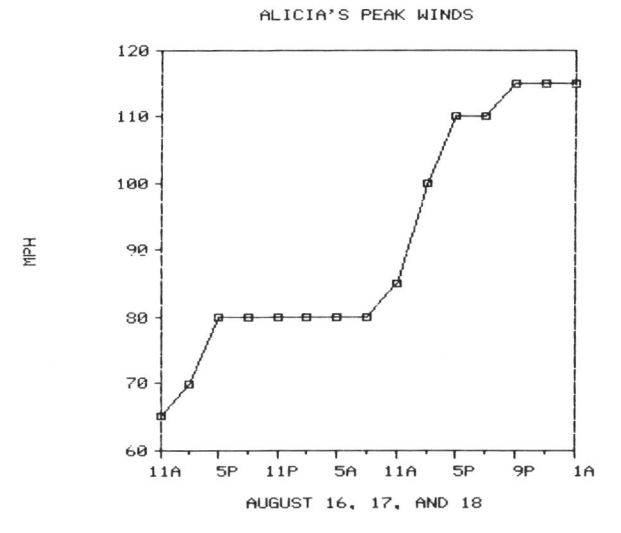

Figure 18.8

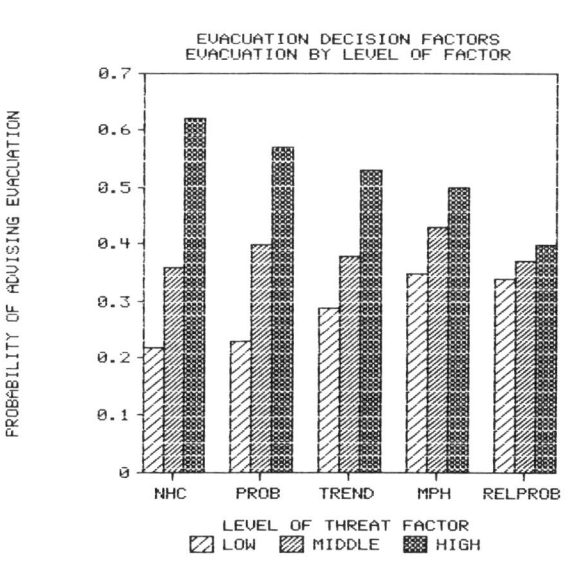

EVACUATION DECISION FACTORS
EVACUATION BY LEVEL OF FACTOR

Figure 18.9

In other examples, Hurricane Celia intensified from 988 mb to 943 mb minimum barometric pressure in its last 14 hours. To illustrate a change in the opposite direction, Carmen went from 935 mb to 980 mb during the last 18 hours.

Probabilities

In 1983, the NHC began issuing information about the real-time (rather than historical) probability of a hurricane striking various coastal sites. The numbers are actually an expression of the error distribution of past position forecasts (Carter, 1983), but the important point for decision makers was that, for the first time, they were being given a quantitative indication of a threatening storm's likelihood of striking their area during the next 24, 36, 48, or 72 hours. They could now compare their probability to that of other locations and ascertain whether their probability had been increasing, decreasing, or remaining about the same over the past few forecasts.

The NHC probabilities can be a valuable tool, but they do not address intensification. Science Applications, Inc. has developed a probability system used by the U.S. Navy that considers intensification likelihood and provides the probability of winds of a specified level for various locations (Jarrell, 1979). Other approaches have attempted to "normalize" the probabilities in one way or another so that they still give some quantitative indication of the risk from location to location, but are no longer interpretable as probabilities (Simpson, 1984).

Current Decision Strategies

To ascertain how emergency management professionals are currently using the sort of information described above, a set of hypothetical hurricane threat scenarios were sent to about 250 local "civil defense" directors along the U.S. Gulf and Atlantic coasts. Responses were obtained from about half the sample, and the findings from the first 100 of those are presented below.

The Threat Scenarios

Twenty-seven hurricane threats were described in the scenarios in terms of five factors (or types of threat information):

1. *Severity of Storm:* In each threat situation the threatening storm either had 95 mph peak winds or 150 mph peak winds
2. *NHC Alert:* In each threat, the National Hurricane Center had either issued a watch, warning, or neither for the storm
3. *NHC Probabilities:* Each threatening storm had either a .10, .30, or .50 probability of striking the respondent's location

4. *Relative Probability:* In each case, the respondent was told whether his probability was higher, about the same, or lower than probabilities at most other locations
5. *Trend in Probabilities:* In each threat situation, the respondent's probabilities had been increasing, remaining stable, or decreasing over the past several hours.

For example, the respondent might be facing a threat in which the hurricane's peak winds were 95 mph, there was no watch or warning in effect, the storm's probability at his location was .10, his probability was lower than most others, and it had been decreasing over the past several hours. The 27 situations were constructed so that the five threat factors were statistically independent of one another, thus creating some unlikely, if not impossible, threat situations. In each case, the emergency management respondent was asked whether he or she would advise evacuation in vulnerable areas in his or her jurisdiction. Local circumstances obviously vary from place to place, but remain the same across the 27 threat situations for each respondent.

Figure 18.9 depicts the amount of weight placed upon each threat factor, averaging over all 100 respondents. There are three bars for each threat factor: the leftmost bar represents the respondent's probability of advising evacuation at the lowest level of the factor (no watch or warning from the NHC, for example); the middle bar indicates the probability of advising evacuation at the middle level of the threat factor (watch from NHC); and the rightmost bar of each set of three gives the evacuation probability at the highest level of the threat factor (warning from NHC). Within a set of three bars, the greater the variation in their height, the greater the weight the respondent was placing on that factor in deciding whether to advise evacuation. The middle bar for MPH (severity of storm) is estimated, having been interpolated from the other two (95 mph and 150 mph).

An alert from the NHC was found to the most important factor, increasing the emergency manager's probability of advising evacuation by 40%, averaging over all the different threat situations. Most of that increase occurred between a watch and a warning. Raw probability information was almost as important, followed by the trend in probabilities. Going from stable probabilities to an increasing trend had a greater effect than going from a decreasing trend to stable probabilities. Comparing the probability to others was least important, accounting for only a six percentage point difference in the likelihood of advising evacuation.

Conclusion

Coastal emergency management officials know that if a severe hurricane strikes their communities, evacuation will be necessary to prevent catastrophic loss of life. They naturally want to avoid unnecessary evacuation, however. Uncertainty in forecasting hurricane track, forward speed, and intensity make it impossible for officials to know exactly when evacuation is or is not necessary. Evacuation is a rare decision for most communities. Therefore, decision makers are often uncertain about how to integrate the information they receive about a threatening storm in order to assess the need for evacuating.

For many years, the National Hurricane Center has effectively made evacuation decisions for coastal communities by issuing warnings for designated reaches of coastline. Emergency management officials have taken the issuance of warnings as their cue to initiate evacuation in their areas. The NHC now disseminates forecasts of future storm positions and intensities, along with "landfall" probabilities to reflect uncertainties in position forecasts. This wealth of information is intended by the NHC to help emergency management officials assess their area's risk and go beyond consideration of warning declarations in making their evacuation decisions.

In hypothetical hurricane threats described to emergency management professionals in the study reported here, a significant number of respondents misused or ignored threat information such as probabilities, trends in probabilities, relative probabilities, and storm severity. Many decision makers are still relying more than they should upon the posting of hurricane warnings. Emergency management professionals in coastal communities need better instruction about the availability of different kinds of threat information and how to develop their own decision strategies to reflect the needs of their constituencies.

References

Baker, E.J. 1983a. *Public Response to Hurricane Probability Forecasts.* NOAA Technical Memorandum to NWS FCST 29. Silver Spring, MD: National Weather Service.

Baker, E.J. 1983b. "The Pro's and Con's of Vertical Evacuation." Paper presented at Fifth National Hurricane Conference, Tampa, Florida.

Carter, T.M. 1983. *Probability of Hurricane/Tropical Storm Conditions: A User's Guide for Local Decision Makers.* Silver Spring, MD: National Weather Service.

Carter, T.M. *Hurricane Graphic Decision System.* Tallahassee, FL: Hazards Management Group, Inc. (computer software), 1984.

Clark, J. 1976. *The Sanibel Report: Formulation of a Comprehensive Plan Based on Natural Systems.* Washington: The Conservation Foundation.

Griffith, D.A. 1979. *Lee County, Florida Hurricane Emergency Evacuation Plan.* Fort Myers: SW Florida Regional Planning Council.

Jarrell, J.D. 1978. *Tropical Cyclone Strike Probability Forecasting.* Final Report to NEPRF. Monterey, CA: Science Applications, Inc.

Ruch, C.A. 1983. *Hurricane Relocation Planning for Brazoria, Galveston, Harris, Fort Bend, and Chambers Counties.* College Station, TX: Sea Grant College

Program.

Simpson, R.A. 1984. "A Risk Analysis and Preparedness System for Use by Coastal Communities." In *Proceedings, Annual Technical Conference on Hurricanes and Tropical Meteorology*. Beacon Hill, MA: American Meteorological Society.

Tampa Bay Regional Planning Council. 1984. *Tampa Bay Regional Hurricane Evacuation Plan Technical Data Report Update*. St. Petersburg, FL: The Council.

Urbanick, T. 1978. *Texas Hurricane Evacuation Study*. College Station, TX: Texas A & M University, Texas Transportation Institute.

STRUCTURAL EVALUATION OF HURRICANE SHELTERS

Christopher P. Jones

Florida Sea Grant Extension Program
Gainesville, Florida

and

Byron Spangler

University of Florida

Introduction

Development on many coastal barriers has increased to such a level that, in the event of a hurricane, it is no longer possible to evacuate everyone within the warning time provided by the National Hurricane Center (typically 12 hours or less). People must either evacuate before hurricane warnings are issued or risk staying behind and exposing themselves to hurricane forces. Many will take that risk.

The problem is made worse by the fact that, while coastal populations have been increasing dramatically in recent years, the incidence of hurricanes striking our shorelines has remained below the historical average. As a result, few people now living along the coast have experienced a hurricane; even fewer have experienced a major hurricane (defined as a Category 3 or higher; see table 19.1). In the event of a storm, these people may not believe that evacuation is necessary until it is too late. Figure 19.1 illustrates this fact for one of the most populated areas in Florida: Pinellas County (the St. Petersburg/Clearwater area on the Gulf Coast). The figure shows that the last major hurricane to strike the area was in 1921, when the population of the area was about 5% of the present level.

TABLE 19.1

SAFFIR/SIMPSON HURRICANE SCALE

Storm Category	Central Pressure (in .Hg)	Winds (mph)	Storm Surge (ft.)*	Damage
1	> 28.94	74 - 95	4 - 5	Minimal
2	28.50 - 28.91	96 - 110	6 - 8	Moderate
3	27.91 - 28.47	111 - 130	9 - 12	Extensive
4	27.17 - 27.88	131 - 155	13 - 18	Extreme
5	< 27.17	> 155	> 18	Catastrophic

*storm surge elevations are averages - they will vary depending on the location.

The authors recently completed a study of hurricane shelters in the Florida Keys (Monroe County) (Spangler and Jones, 1984). Although the population at risk there numbers in the tens of thousands, rather than hundreds of thousands, all of those residents must evacuate over a single, two-lane highway. Many people in the lower Keys have indicated that they will not attempt to leave when a hurricane strikes but will seek refuge in buildings on higher ground.

Unfortunately, the safety of those who do not evacuate from the Florida Keys or from other coastal areas cannot be guaranteed. Despite this, the local government must take steps to minimize the risk to those who remain. One option is to shelter people in buildings thought capable of withstanding hurricane forces. The best alternative is to place people in fully engineered, well-built structures located away from the influence of flooding and waves. This may not always be possible, especially in low-lying areas. The next best alternative is to shelter people in the upper stories of buildings in which the lower floors may be subject to flooding. This is termed "vertical evacuation."

There are many technical and policy problems associated with vertical evacuation. First, does designating vertical evacuation shelters encourage people to remain rather than to evacuate? This question has yet to be answered. Other problems and concerns are: What is an acceptable level of risk to shelter occupants? When a building is designated as a vertical evacuation shelter, what are the liabilities of the building owner, the designer, the contractor, the local government, and the inspector rating the building? What should be done in areas where complete evacuation is not possible, yet where existing structures offer shelter only during minor storms? What is the best method of evaluating potential shelters?

There is a small but growing body of literature dealing with some of these problems (Salmon, 1984; Saffir, 1984; Rogers, 1984). There are also many reports dealing specifically with wind damage to buildings that are useful in evaluating shelters (Minor and Mehta, 1977; Mehta, McDonald, and Smith, 1981; Mehta, Minor, and Reinhold, 1983).

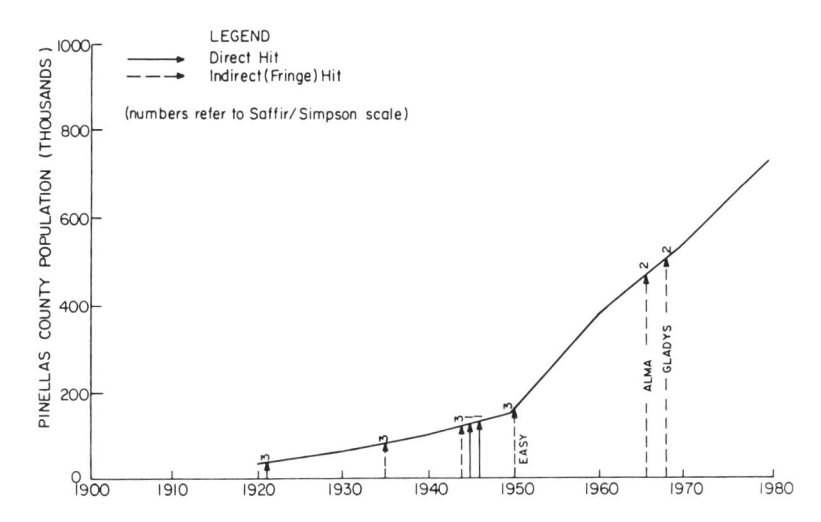

Figure 19.1. Hurricane Experience Levels of Pinellas County Population (adapted from Hebert and Taylor, 1975, p. 51)

It is not possible to make an exact determination of the level of protection that an existing building can provide during hurricanes. Uncertainties relating to the design and construction of the building and the storm forces that act on it tend to limit the accuracy of the evaluation. Hence, *vertical evacuation should be used only as a last resort.*

General Procedure for Shelter Evaluation

The resistance of a shelter to hurricane forces must be determined as accurately as possible. If the resistance is underestimated, the use of the shelter under some conditions will be lost. If the resistance is overestimated, the building may sustain structural damage and the occupants may be injured or killed. Two types of factors limit the accuracy of evaluation of an existing building: (1) those associated with the building itself and (2) those associated with the hurricane forces that will act on the building.

Evaluation of an existing building requires design calculations, complete plans and specifications, information on materials used during construction, and a record of building modifications and maintenance. These are often difficult to assemble.

It is also difficult to predict accurately storm forces at the shelter site for a hurricane of a particular intensity. Small-scale spatial and temporal variations in the hurricane wind field cannot be predicted precisely. Storm surge elevations can be predicted approximately with numerical models, but actual elevations will fluctuate due to variations in wind and local topography.

Uncertainty in the evaluation can be reduced, however, by collecting as much information about the building and site as possible, carefully inspecting them, obtaining the best available predictions of hurricane forces at the building site, and performing the appropriate structural analyses. The performance of similar buildings during past hurricanes should be studied as well.

Other investigators have developed procedures for evaluating existing buildings, but these have not taken into account the effects of flooding, currents, and waves. Culver et al. (1975) presented a methodology for the survey and evaluation of existing buildings subject to earthquake, hurricane wind, and tornado forces. Mehta, McDonald, and Smith (1981) offered a methodology for predicting wind damage to buildings. Both have presented a subjective procedure based upon an on-site inspection and more detailed analyses of building response to storm forces. The Monroe County shelter study involved building and site inspections, with some structural analyses, taking into account the effects of wind, flooding, currents, waves, and debris (Spangler and Jones, 1984).

The following list outlines a suggested general procedure for evaluating hurricane shelters:

1. *Identify Potential Shelters.* This should be based on location, elevation, type of construction, etc.
2. *Collect Information.* Obtain plans (as-builts, where possible), specifications, building modification and maintenance records; locate the designer and contractor; obtain flood hazard data.
3. *Inspect the Building.* Check all structural elements and connections, where possible; note any deviations from the plans and any defects or problems; obtain samples of materials and perform tests on building components, as required; photograph and document the building and its condition.
4. *Inspect the Building Site.* Check for exposure to wind and proximity to water; photograph and document adjacent structures or vegetation that may shield or damage the building.
5. *Analyze all Information.* The procedure and effort will depend upon the amount of information available and the complexity and condition of the building.
6. *Rate the Building.* Determine the highest category storm (see table 19.1) that the building could withstand, taking into consideration the effects of wind and water; note any special precautions or repairs that must be made before the building can be used as a shelter.

The authors have developed a shelter summary form that is useful when collecting information and inspecting the shelter and site. The reader is referred to the complete report for a detailed discussion of the information to be collected (Spangler and Jones, 1984). This study was concerned with the structural resistance of shelters, i.e., with the survival of the occupants rather than with their comfort. Hence, there are several items that are not included on the form that are necessary from the standpoint of overall shelter suitability, such as: available space, emergency power, kitchen facilities, restroom facilities, emergency supplies, etc.

Rating Shelters

Once the building and site information are collected, the next step is to determine the hurricane forces that will act on the structure, namely, wind forces and water forces. The former include positive and negative wind pressures and the effects of wind-driven missiles. The latter include flooding, hydrostatic forces (including flotation), hydrodynamic forces (including scour), breaking wave forces, and the effects of water-borne debris. Estimates of wind and water forces should be tied to the Saffir/Simpson scale, since evacuation plans and decisions are usually based upon the category of the hurricane approaching the area (or the category expected at the time of landfall).

Wind speeds, storm tide still water levels, and breaking waves should be taken as the maximum, that can occur at the shelter site for a given category storm (e.g., for a Category 2 storm, analyze for 110 mph winds, etc.). This approach is necessary since the exact point of landfall of an approaching storm cannot be predicted. It must be assumed that the shelter will be subject to the most severe conditions the storm can generate.

The resistance of a shelter can be defined as the highest category storm for which it can withstand both wind and water forces acting concurrently without jeopardizing the occupants. Note that this does not exclude the possibility that there may be minor damage to the shelter, as long as the major structural elements (foundation and structural frame) and protective elements (walls, openings, roof) are intact.

The use of a shelter will be limited by the lesser of its resistance to wind forces and water forces. For instance, a building safe from flooding, current, and wave effects during a Category 4 storm will be of no use during those conditions if it can withstand only Category 2 wind forces; a single-story building that can withstand Category 3 wind forces will be of no use if it is flooded during a Category 1 storm.

In general, a building should not be used as a shelter during a given category storm if any one of the following can occur:

- For single-story shelters:
  1. If the building cannot resist anticipated wind forces
  2. If water can rise above the level of the floor

3. If settlement of the foundation or floor will occur as a result of scour by waves or currents
4. If the building cannot withstand anticipated breaking wave forces
5. If the building cannot resist flotation, sliding, or overturning.
- For multi-story shelters:
  1. If the building cannot resist anticipated wind forces
  2. If all floors of the building are flooded
  3. If settlement of the foundation will occur as a result of scour by waves or currents
  4. If the structural frame cannot withstand anticipated breaking wave forces (acting either directly on the frame or transferred to the frame by lower story walls)
  5. If the building cannot resist flotation, sliding or overturning
  6. If the structural frame cannot withstand battering by floating debris.

Upper stories of multi-story shelters are considered adequate if they do not flood and if the shelter can resist the above-mentioned forces (even if lower stories flood).

The first step in evaluating a shelter's resistance to water forces is to determine the *total* water level at the site for each category storm. This includes the storm tide still water level (swl) plus the height of any waves above the swl. The storm tide is found by taking the storm surge (the output of the SPLASH and SLOSH computer models) and adding the contributions of the astronomical tide and wave setup. Wave heights should then be added using the methodology developed for the Federal Emergency Management Agency by the National Academy of Sciences (FEMA, 1981).

Given the total water level at the shelter site, the next steps are to compute hydrostatic forces (both lateral and vertical) and hydrodynamic forces. Guidance may be obtained in several references (Texas Coastal and Marine Council, 1981, p. IV-43; Colorado Water Conservation Board, 1983, ch. 6). Breaking wave forces should then be calculated. Breaking wave pressures can reach hundreds or thousands of pounds per square foot (other loads may be on the order of tens of pounds per square foot). These breaking wave pressures, although intense, are of a very short duration and are rated as impact loads in the structural analysis. Guidance for calculating wave forces can be obtained in a recent paper (Kirkoz, 1982) and in the shore protection manual (U.S. Army Corps of Engineers, 1984, chapter 7). While these methods may result in conservative estimates of breaking wave forces, the authors think this is justified, given the fact that people will occupy the building during storms.

Finally, the evaluation of a shelter must take into consideration the effects of debris carried by wind or water striking the building. While it is almost impossible to predict exactly the debris that may strike a shelter during a hurricane, a rough estimate may be developed from observance of other buildings and objects (towers, tanks, signs, trees, vehicles, etc.) in the immediate vicinity.

Monroe County Study

The authors evaluated 31 designated or potential hurricane shelters in Monroe County (fig. 19.2). Many of the shelters have not experienced hurricane conditions. Figure 19.3 shows the tracks, dates, and intensities of hurricanes that have affected the Keys since 1900. Only two of the shelters in the upper Keys and two of the shelters in the lower Keys have experienced a major storm. It is likely that shelters in other coastal areas have not been exposed to hurricane conditions very often either.

Inspections revealed that only a few of the shelters had adequate shutters; most had no shutters at all (Monroe County is in the process of installing shutters now). In many instances, buildings could not be recommended for use as shelters until repairs are made. Even if shutters are installed and necessary repairs are undertaken, very few of the 31 buildings would be capable of providing shelter during a major hurricane. The survey resulted in the following designations: Four should not be used as shelters under any circumstances, three could be used during Category 1 storms, 14 could be used during Category 2 storms, eight could be used during Category 3 storms, one could be used during Category 4 storms, and one could be used during Category 5 storms.

The results are not surprising when one considers the fact that most of the buildings in the region have been designed to withstand 100 to 120 mph winds (Category 2 and 3 winds) and that many have not been designed to withstand scour and wave forces. The results are also consistent with the behavior of similar buildings in other areas during hurricanes. For example, a damage survey after Hurricane Frederic revealed that schools, churches, public buildings, and hospitals—the types of buildings typically used as shelters—sustained extensive damage (U.S. Army Corps of Engineers, 1981, pp. 144-151).

The Monroe study and post-storm damage surveys in other areas point out several items that should be inspected and analyzed carefully during any shelter evaluation. These include:

- exposure
- scour potential
- foundation
- exterior walls
- openings
- shutters
- connections
- roof
- roof drainage
- maintenance

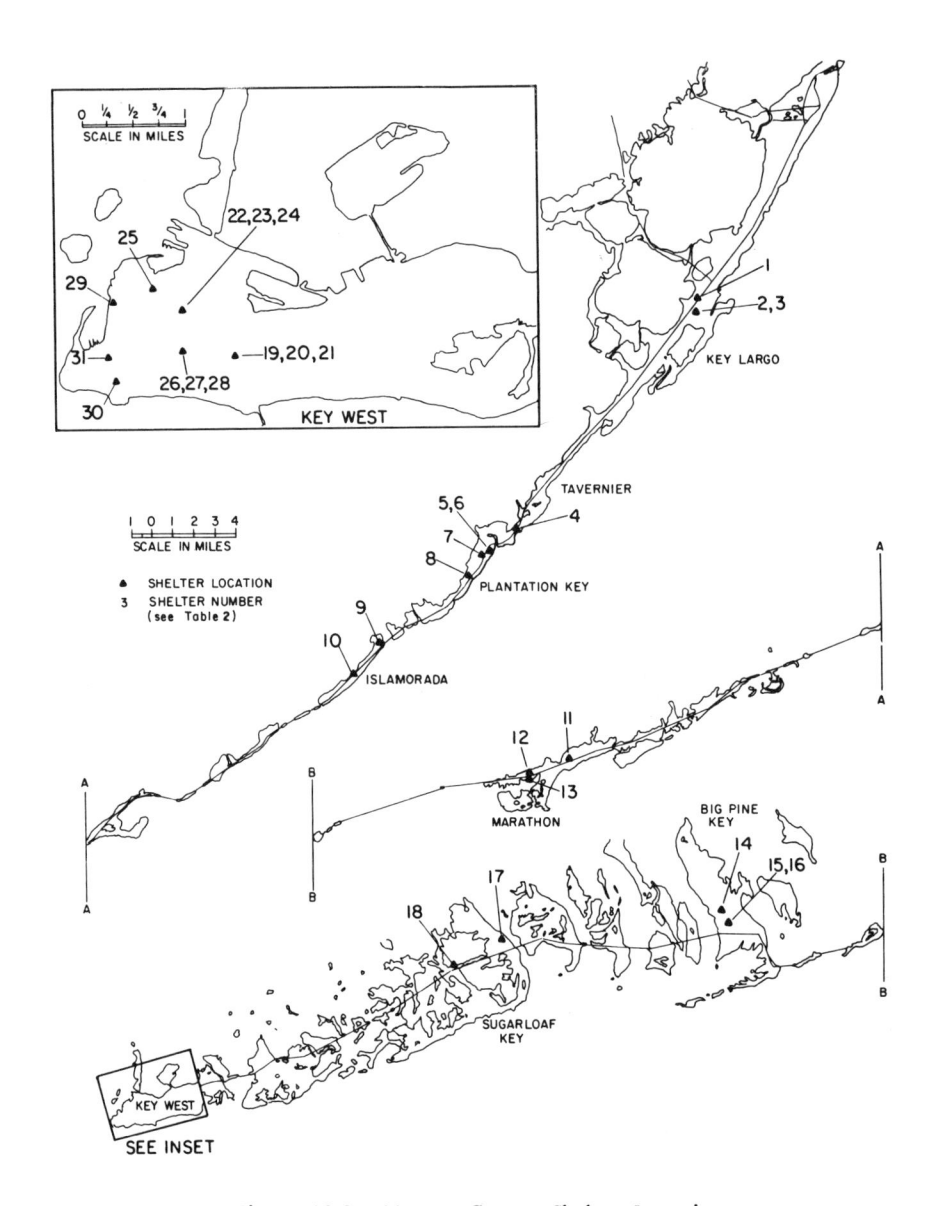

Figure 19.2. Monroe County Shelter Locations

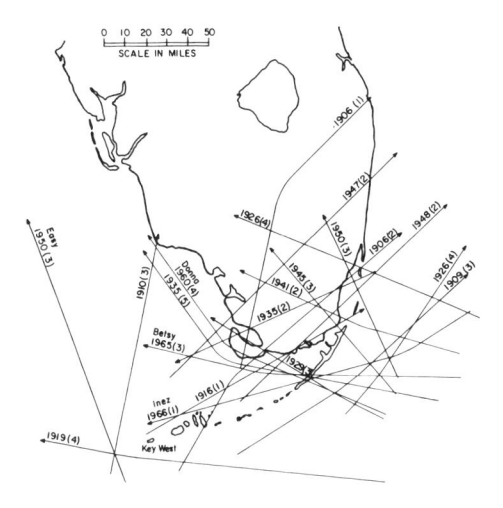

Figure 19.3. Selected Hurricanes affecting Monroe County since 1900 (Post, Buckley, Schuh, and Jernigan, 1983, p. 16)

Summary

This paper has outlined a procedure for the structural evaluation of hurricane shelters, including vertical evacuation shelters. The reader is referred to the complete report for more details (Spangler and Jones, 1984). The procedure, which takes into account wind, flooding, current, wave, and debris forces, was developed during a study of hurricane shelters in Monroe County, Florida. The study revealed that very few of the buildings evaluated are capable of providing shelter during a major hurricane. It is expected (on the basis of hurricane damage studies in other areas) that similar results would be found almost anywhere along our coastline.

Acknowledgments

The authors would like to thank the Florida Sea Grant College and Monroe County for their financial support of this work. In addition, the assistance of Monroe County Civil Defense, the U.S. Army Corps of Engineers, the U.S. Navy and the National Hurricane Center is acknowledged.

References

Colorado Water Conservation Board. 1983. *Colorado Floodproofing Manual.* Denver: Colorado Department of Natural Resources.

Culver, C.G., H.S. Lew, G.C. Hart, and C.W. Pinkham. 1975. "Natural Hazards Evaluation of Existing Buildings." *Building Science Series 61.* Washington: U.S. Department of Commerce, National Bureau of Standards.

Federal Emergency Management Agency. 1981. "Ways of Estimating Wave Heights in Coastal High Hazard Areas in the Atlantic and Gulf Coast Regions." *Resource Document TD-3.* Washington: FEMA.

Hebert, P.J., and G. Taylor. 1975. *Hurricane Experience Levels of Coastal County Populations - Texas to Maine.* Washington: U.S. Department of Commerce, National Weather Service.

Kirkoz, M.S. 1982. "Shock Pressures of Breaking Waves on Vertical Walls." *Journal of the Waterway, Port, Coastal and Ocean Division - American Society of Civil Engineers.* Vol. 108, no. WW1, pp. 81-95.

Mehta, K.C., J.R. McDonald, and D.A. Smith. 1981. "Procedure for Predicting Wind Damage to Buildings." *Journal of the Structural Division - American Society of Civil Engineers.* Vol. 107, no. ST11, pp. 2089-2096.

Mehta, K.C., J.E. Minor, and T.A. Reinhold. 1983. "Wind Speed—Damage Correlation in Hurricane Frederic." *Journal of the Structural Division—American Society of Civil Engineers.* Vol. 109, no. ST1, pp. 37-49.

Minor, J.E., and K.C. Mehta. 1977. "Wind Damage Observations and Implications." *Journal of the Structural Division - American Society of Civil Engineers.* Vol. 105, No. ST11, pp. 2279-2291.

Post, Buckley, Schuh and Jernigan, Inc. and U.S. Army Corps of Engineers, Jacksonville District. 1983. *Lower Southeast Florida Hurricane Evacuation Study.* Technical Data Report.

Rogers, B.A. 1984. "A Hurricane Refuge of Last Resort for a Barrier Island." *Proceedings of the Third Conference on Meteorology of the Coastal Zone.* Boston: American Meteorological Society. pp. 70-72.

Saffir, H.S. 1984. "Structural Integrity of Potential Vertical Evacuation Shelters." *Proceedings of the Third Conference on Meteorology of the Coastal Zone.* Boston: American Meteorological Society. pp. 57-60.

Salmon, J.D. 1984. "Vertical Evacuation in Hurricanes: An Urgent Policy Problem for Coastal Managers." *Coastal Zone Management Journal.* Vol. 12, nos. 2 and 3.

Spangler, B.D. and C.P. Jones. 1984. "Evaluation of Existing and Potential Hurricane Shelters." *Florida Sea Grant Report 68.* Gainesville, Fla.

Texas Coastal and Marine Council. 1981. *Model Minimum Hurricane Resistant Building Standards for the Texas Gulf Coast.* Austin.

U.S. Army Corps of Engineers. 1981. *Hurricane Frederic Post Disaster Report.* Mobile: Mobile District.

U.S. Army Corps of Engineers. 1984. *Shore Protection Manual.* Vicksburg: Coastal Engineering Research Center.

HURRICANE ALICIA AND THE GALVESTON EXPERIENCE

James M. McCloy

Texas A & M University at Galveston

and

Stephen N. Huffman

City Manager, Galveston, Texas

Introduction

At approximately 3:00 a.m. on August 18, 1983, hurricane Alicia made landfall on the west of Galveston Island, Texas. Alicia had come into being as a tropical storm in an area of disturbed weather over the northern Gulf of Mexico three days earlier, on August 15, and had drifted slowly westward. It became a minimal hurricane (Saffir/Simpson scale, Category 1), on August 16, remained weak (85 mph at 11:00 a.m., August 17), then intensified very rapidly to a Category 3 hurricane with 115 mph winds (3:00 p.m., August 17). After crossing Galveston Island, it severely battered the Houston metropolitan area before moving onto the Great Plains (Subcommittee on Natural Resources, 1984, p. 8-9). The National Weather Service, 18 hours before landfall, had stated that the probability of the hurricane coming within 65 miles of Galveston was 51%. In all, the hurricane caused 11 deaths and estimated property damage of $1.7 billion (Morrison, Norton, and Simmons, 1984, p. 8). (There were no deaths on Galveston Island, however.)

Galveston Island is a low-lying barrier island, aligned northeast to southwest, and approximately 28 miles long. The average shoreline elevation is five feet, and the highest natural land—beach ridges on the bay side of the island—is only about ten feet above Mean Sea Level (MSL). In 1900, a hurricane estimated to be a

Category 4 or 5 struck Galveston, leaving more than 6,000 dead. It was the largest single loss-of-life disaster in United States history (Subcommittee on Natural Resources, 1984, p. 32).

To prevent another hurricane catastrophe, Galveston began to construct a seawall in 1902 at the eastern end of the island and extended it westward in segments for a distance of 10.3 miles, finally completing the structure in 1962. To further mitigate flooding from the bay side during storms, the grade of the island was raised to the crest of the seawall, 17 feet above mean low water, with a gentle slope to the bay. The grade-raising project covered only the eastern half of the island behind the seawall, which to this day is the most heavily populated portion of the island. The seawall has yet to be breached by a storm (Davis, McCloy, and Craig, in press). During Hurricane Alicia, the two-thirds of the island west of the seawall was subjected to a storm surge of six to ten feet (Region 6 Interagency Hazard Mitigation Team, 1983a, p. 3), and the Texas Land Commission recorded a shoreline retreat for the unprotected west end beaches of 75-100 feet, with concomitant vertical erosion of four to five feet (Barron Publications, 1983, p. 7).

The 62,000 residents of Galveston Island have three possible evacuation routes: ferryboat transit to Bolivar Peninsula on the east, a bridge at the west end to the mainland, and a three-lane causeway to the mainland. Both the ferryboat and the bridge to the west become impassable when the water reaches four feet above mean sea level as the ferries must secure for the storm and the beach road to the west of the bridge floods. There was no order to evacuate the island during Hurricane Alicia, although the people on the west end were advised to go to high ground on either the island or the mainland. It is estimated that approximately 10% of the population evacuated prior to arrival of the full force of the hurricane (Martin, 1984, p. 11).

Damage to residential and commercial buildings on the west end was severe (table 20.1). Those on the ocean sustained storm surge and wave damage, as well as wind damage. Buildings inland from the strand line were, in some instances, demolished, possibly because of small tornados. There is some debate as to whether or not there actually were tornados; some feel that these summer cottages were just not well constructed.

Post-Hurricane Management

Utilities

Restoration of utilities was one of the first priorities after the storm. According to the Houston Lighting and Power Company, all transmission lines to Galveston Island were knocked out at 5:30 a.m. on August 18, and in the Houston/Galveston area approximately 750,000 customers were without electric service. The first transmission line to Galveston was back in service by 2:30 a.m. the following day. By 3:30 a.m., service had been restored to three of the four Galveston substations,

TABLE 20.1

ESTIMATED STORM DAMAGE - GALVESTON ISLAND

Damage Category	Cost (Millions of Dollars)
Commercial Damage	$ 355.0
Residential damage	314.0
City public property damage	9.4
University of Texas Medical Branch	8.0
	686.4

Housing Units by Type	No. Destroyed	No. With Major Damage
Single-family homes	1,062	6,750
Mobile homes	764	443
Apartment Units	262	785
	2,088	7,978

SOURCE: Subcommittee on Natural Resources, 1984, p. 78-79.

two hospitals, the City Hall, the main fire station, two water plants, and a portion of the downtown area. By nightfall, August 19, service was restored to about 10% of Galveston Island customers behind the seawall. Work was concentrated on major city water and sewer plants, critical customers, grocery stores, service stations, civil defense, Red Cross, and major feeder circuits. Most residential units had power within five days (Subcommittee on Natural Resources, 1984, p. 187).

Law and Order

City officials requested that the Governor send National Guard troops and Department of Public Safety troopers to control traffic onto the island. They were assigned to keep nonresident sightseers off the island, to restrict access to the west end of the island to residents of that area, and to assist police in deterring looting and in enforcing the 8:00 p.m. to 6:00 a.m. curfew. The Chief of Police feels strongly that the proper actions were taken in that there were only 18 more burglaries in August 1983 than in August 1982; traffic control which limited access to island residents only was exceptional and well received by the residents; less than 40 people were arrested for looting because of the strong police and military forces on patrol; and during curfew the island was in fact tranquil (Police Chief Eddie Barr, personal communication, Dec. 2, 1985).

Disaster Relief Center

The Federal Emergency Management Agency (FEMA) sent staff to Galveston two days after the storm to set up a Federal Disaster Relief Center for the citizens of Galveston County and to assist officials of the city. The FEMA contingent was very well organized and extremely helpful. The City provided the Civic Center Auditorium for the Federal Disaster Relief Center. City authorities feel that some reimbursement of extraordinary expenses in regard to providing the Disaster Relief Center should be included in the 75% reimbursement by FEMA.

Debris Clearance

Federal officials and the U.S. Army Corps of Engineers were very helpful in assisting the city with developing bid specifications and contract awards for the clearance of debris. The specifications took time to develop and caused some delay in clearing the debris. In the future, it is recommended that FEMA provide at least the guidelines to hurricane-prone cities in advance, so that the process of developing bid specifications, review, and contract awards will be expedited. If criteria regarding the debris cleanup bid specifications should change at any time during the year, FEMA should notify the local governments accordingly. This would allow the city to have the preparation of bids and awards well underway when FEMA officials arrive on the scene of the disaster.

Reimbursement by FEMA

The City of Galveston was faced with a serious budget shortfall, which made the 25% local contribution to disaster recovery extremely difficult. In January of 1979, local residents imposed budget limitations on the city, making it almost impossible to keep up with inflation. On Galveston Island, fully 45% of the property is tax-exempt. City officials have therefore recommended to the federal government that the FEMA reimbursement be raised from 75% to 95% in certain cases—possibly by constructing a formula based upon the percentage of federal, state, county, and local tax-exempt properties.

Recovery Task Force

Approximately five days after Hurricane Alicia hit the City of Galveston, the City Council appointed a 20-person Recovery Task Force. The committee was divided into subcommittees to cover such areas as a construction moratorium and controls, price freezing, temporary housing, utility restoration, health and emergency medical, west end blockades, communications and civil defense, emergency shelters, evacuation and return, insurance settlements and maintenance, relief needs, and return to normalcy. This task force made recommendations for changes to the building code and zoning areas which were later adopted by the

City Council. The task force was assisted by a federal interagency "hazard mitigation team" under FEMA's direction.

The long-range responsibility of the Recovery Task Force is to develop hazard mitigation procedures to guide future development. An effective approach would be to prepare a growth management system based on an analysis of the area's carrying capacity (see Brower and Beatley paper in this volume). This development management system should take into consideration the following factors: (1) limited water supply, (2) ability of soil to absorb wastes, (3) capabilities of traffic and evacuation, (4) availability of land, (5) environmentally sensitive nature of the island, and (6) limitations on multi-hazard areas.

It is suggested that FEMA hold training sessions for the federal agencies that normally support the Damage Survey Report (DSR) process. It is apparent that many of these are not fully aware that mitigation measures can be proposed in addition to what is normally considered adequate to restore structures to pre-disaster conditions. In many cases, mitigating measures (e.g., flood-proofing) can comprise up to 15% of eligible DSR work.

The City of Galveston recommends that provision for technical assistance be given to cities to study the recommendations made by the hazard mitigation team. Once the mitigation team has identified problem areas, local governments often lack the expertise needed to explore solutions to the problems. Often they also lack the financial resources to hire outside consultants to develop solutions to the problems.

Rehabilitation

House Repair and Reconstruction

On the east end of the island, behind the seawall, there was surprisingly little major structural damage to residential units. However, some commercial units facing the Gulf of Mexico did suffer major structural damage. This damage was inflicted mostly on new condominium/apartment complexes, new housing, and the buildings to the west, beyond the seawall. Quite a bit of roofing was lost that had been attached only by staple guns. The first row of oceanfront homes was highly damaged by both the storm surge and waves; the second row fared much better; and up to 200 feet inland the damage was scattered. The most severe damage was possibly the result of tornados.

Approximately 50 property owners are being sued by the State of Texas as a result of the hurricane. The Texas Open Beaches law declares all areas seaward of the vegetation to be public. Erosion of the shoreline moved the leading edge of the vegetation landward a considerable distance, thereby placing some houses on the public beach. Therefore, the Texas Attorney General deems that houses left on the beach forward of the erosional hurricane vegetation line are on state property. People were allowed to rebuild these houses if damage was 50% or less.

However, some owners with more than 50% damage chose to rebuild and some even erected bulkheads and brought in sand on this "public" beach without permit or permission (Region 6 Interagency Hazard Mitigation Team, 1983b, pp. 3-7, 12, 14).

The depth of flooding on the west end of the island was two to four feet above the existing ground elevation. The beachfront homes that were destroyed were thought to have not had piling embedment to withstand the combined wind and water forces. Generally, it was concluded that the property damage resulting from the hurricane surge and associated wave action was relatively minor when compared to the extensive property damage caused by the accompanying wind and wind driven rain. Actual scour depths of three to five feet are typically accompanied by an area of sand liquification below the scour level, which contributes to the instability of the foundation. Therefore, it was recommended that piles should be embedded to ten feet below MSL depth. The majority of the pile foundations were tied together by concrete slabs at the ground elevation. Many of these slabs failed as a result of undermining and the effects of the storm surge. Most of the structural damage on the island was caused by the wind and was a result of inadequate construction of the wall and roof systems. A number of homes that should have breakaway walls at ground level instead had solid wall construction, which could have contributed to the destruction of some pilings when broken up by the surge and waves. The quality of the construction on the west end of the island and the number of wind failures indicates that some aspects of the local building code should be improved. Ironically, a stricter building code went into effect and three new building inspectors were approved on August 4, 1983, just two weeks before the occurrence of Hurricane Alicia.

Conclusion

Though struck by a Category 3 hurricane, Galveston Island and its residents did not suffer the effects that a similar storm might inflict on other coastal communities. The reasons are diverse and many, but here are the major points. The hurricane quickly developed from a Category 1 to a Category 3 and then made almost immediate landfall, which generated only a modest storm surge. The vast majority of people were protected by a 17-foot seawall on the eastern third of the island. Public safety officials and the National Guard very effectively controlled island access and strictly enforced a curfew. The utility companies did an outstanding job in restoring services. The citizens worked together after the hurricane to assist one another. Finally, the citizen Recovery Task Force provided leadership and hazard mitigation analysis.

The hazard mitigation analysis identified the need to update the emergency management plan and the Emergency Operations Center. As city funds were unavailable, the citizens of Galveston contributed approximately $28,000 to upgrade the communications system and the auxiliary power generators of the

Emergency Operations Center. During the three 1985 hurricanes in the Gulf of Mexico, the Emergency Operations Center plan and equipment operated effectively under actual emergency conditions.

References

Davis, D.W., J.M. McCloy, and A.K. Craig. "Man's Response to Coastal Change in the Northern Gulf of Mexico." In *Coasts of the World*. Ed. K. Ruddel. Paris: International Geographical Union, in press.

Barron Publications, Inc. 1983. *Hurricane Alicia - Thursday, August 18, 1983*. Lubbock, Texas: Barron Publications, Inc.

Martin, N. 1984. "Before the Wind." *Texas Shores*. Vol. 17, no. 2 (Summer). Texas A&M Sea Grant Program, pp. 7-13.

Morrison, R.E., C.F. Norton, and M.M. Simmons. 1984. *Summary of Hearings on Hurricane Alicia*. Washington: Congressional Research Service, Library of Congress.

Personal Communication, Police Chief Eddie Barr, City of Galveston, Dec. 2, 1985.

Region 6 Interagency Hazard Mitigation Team, 1983a. *Interagency Flood Hazard Mitigation Report*. FEMA-689-DR, Appendices A-E. Washington: Federal Emergency Management Agency.

Region 6 Interagency Hazard Mitigation Team, 1983b. *Interagency Post-Flood Recovery Progress Report*. FEMA-689-DR, Appendices A-D. Washington: Federal Emergency Management Agency.

Subcommittee on Natural Resources, Agriculture Research and Environment of the Committee on Science and Technology and the Subcommittee on Water Resources of the Committee on Public Works and Transportation, U.S. House of Representatives. 1984. *Hurricane Alicia - Prediction, Damage and Recovery*. Washington: U.S. Government Printing Office.

REDUCING THE PSYCHOSOCIAL TRAUMA OF A HURRICANE

Alan P. Chesney

University of Texas Medical Branch
Galveston, Texas

Introduction

This paper discusses the human aspects of the impact of a hurricane on a community. It also presents ways in which communities and particularly employers might respond to the aftermath of a hurricane. Comments are based on the author's direct experience with natural disasters, primarily with Hurricane Alicia, which struck Galveston Island, Texas on August 18, 1983, and with the literature on disasters and disaster interventions.

Literature on Disaster Stress

Disasters have historically been related to a wide variety of human problems, including physical and emotional disorders (Melick, 1978; Titchner and Kapp, 1976), depression (Ollendick and Hoffman, 1982), anxiety (Kinston and Rosser, 1974), hypertension (Logue and Hansen, 1980), accidents (Kliman, 1976), and alcohol or drug abuse (Kliman, 1976). However, it is apparent that emotional and physical illnesses do not necessarily result from having experienced a disaster. There appears to be ample evidence that low incidence of emotional disturbance may occur after a disaster (Bates, 1963; Quarentelli and Dynes, 1977; and Hall and Landredth, 1975). Many of the studies on disasters do not give specific or detailed information about the nature of the intervention process. Rather, they are epidemiologic studies of the occurrence of an illness. Therefore, statements about why some disasters produce increased human problems and others do not are speculative. More recently, research has been performed which demonstrates the importance of social support in mediating the effects of natural disasters on people. Research after the Three Mile Island accident in Pennsylvania indicated that the high stress prompted by disaster results in dramatic physiological changes, which in turn may cause behavioral changes. However, in high stress situations where social support is available, there appears to be decreased physical arousal and therefore decreased aberrant behavior.

A second point which emerges from the literature is that early community intervention is an essential ingredient of successful intervention. This point is well illustrated by the Corning Flood Project. The coordinator of the project (as a consultant to Corning Glass) considered the City of Corning to be a model community that

> used a disaster as an opportunity to reorder many of its priorities and to reconsider many of its options in a way which promoted growth and healthy adaptation. Some of the successes of this rebuilding can be attributed to the availability of its citizens to provide psychological first aid through care givers trained for this project under the sponsorship of responsible and respected local groups. Equally essential were the efforts to facilitate community movement from a passive, victimized, transiently regressed position to an act of healthy mastery. (Kliman, 1976).

Since 1972, Ms. Kliman has traveled throughout the world to help communities develop intervention strategies for disasters. Upon her visit to Galveston, Texas after Hurricane Alicia, she urged that the University of Texas Medical Branch establish an inservice training program for its staff. What she suggested was not elaborate, expensive, or time-consuming. It was merely a workshop in which departments would discuss disaster theory and management. This would provide an ongoing and self-sustaining peer support system that would utilize skills that are as applicable to daily work as they are to a crisis situation.

In deciding what type of intervention to make after Hurricane Alicia, it became apparent that it was necessary to consider the various phases of the post-impact situation. Frederick (1978) has identified three stages that occur after a disaster. The first stage is the "honeymoon stage," which can last from one to six weeks. The honeymoon phase is characterized by the exhilaration of having survived the disaster. It is a feeling that this problem can be conquered and resolved and that any other problems that result from the hurricane or disaster will be managed. The next stage is "disillusionment" and can occur from two months to two years after the hurricane. This stage is characterized by anger, fatigue, resentment, irritability, and a feeling of having been let down or betrayed. Much of the anger and resentment is directed toward others, such as government agencies or community leaders. The end of the disillusionment phase is often characterized by extreme physical and psychological exhaustion. A third phase is a "reconstruction phase," which usually occurs up to two years after the disaster. The essential ingredient in the reconstruction phase is that the individual begins to assume responsibility for problem solving. The length of each stage may depend on the nature of the disaster, existing relief efforts and support services, and an individual's coping skills. One of the goals of a support system should be to help people work through the stages more quickly, to minimize the impact of the hurricane.

Berren, Beigel, and Barker (1982) have suggested a five-dimensional typology for disasters. They suggest that this typology is important for deciding what kinds of intervention techniques are most appropriate. The five dimensions are: control over future impact of the disaster, the duration of the disaster, whether the disaster is natural or manmade, the degree of personal impact of the disaster, and the potential for the disaster to reoccur.

In a hurricane situation, the disaster may be described as being of relatively short duration, having high personal impact, including loss of life and property, having a high potential for reoccurrence over a long period of time, and a future impact that can be minimized through community action. Therefore, priority should be placed on educational programs and, particularly, dissemination of information after the hurricane. Community recovery programs should emphasize the prevention of physical and emotional problems. It is not just a matter of treating the victims of hurricane, but of disseminating information and educational materials to the community as a whole. In some cases there is a need for psychotherapy, depending on the degree of personal impact, but the vast majority of a community may be served by providing indirect services.

In summary, the literature seems to support the idea of establishing social support groups, primarily peer groups, which provide both information and emotional support to disaster victims. The development of these support groups should occur within the context of a total community recovery effort. The groups should be planned in such a way that they can be activated within a few days of a disaster.

The Galveston Experience

The University of Texas Medical Branch (UTMB) is the largest employer in Galveston, with an employee population numbering 7,500. At least one member in every four households in Galveston is employed at the medical branch. The Employee Assistance Program (EAP) assesses and treats employees who are having personal problems that are affecting their job performance. In the two years the EAP has been in existence, it has worked with over 500 employees. Employees come to the program with financial or legal problems, marital or family problems, and job-related problems.

Immediately after the occurrence of Hurricane Alicia, the kinds of job problems that began to appear at the medical branch were absenteeism, accidents on the job, poor work performance, and a number of terminations that were directly related to the hurricane. In September 1983, there were more on-the-job accidents at UTMB than for the months of October, November, and December 1983 combined. There was also a substantial increase in absenteeism, although much of it was undocumented.

The personal troubles with which people came to the EAP shortly after the hurricane can be categorized into four groups. The first and most frequent group included financial and/or legal difficulties, such as tenant rights, sources of financial aid, and loan information. The second group involved interpersonal problems which developed as a result of changed housing situations. For example, there were married couples who were no longer living together because they had to find separate housing, or multigenerational families who were forced to live together because one or another house was uninhabitable. These interpersonal problems may have been going on in these families for years, and the hurricane acted as a catalyst to raise these problems to a level such that they could no longer be ignored. The third problem area involved a combination of anxiety and depression, characterized by an inability to sleep, hyperactivity, and constant worry about how things were going to be accomplished. Feelings of being unable to cope with the situation may also have resulted in loss of appetite or dramatic mood changes. The fourth group involved alcohol or drug abuse, which occurred as much as a year and a half after the hurricane. As a result of this disaster and the stress that it created, old coping mechanisms that included alcohol and/or drugs began to be used and slowly escalated into a problem. Even though this behavior may have occurred right after the hurricane, it did not surface as a problem until six months to a year or more later.

It is important to note that there were no reported cases of extreme psychopathology either immediately after the hurricane or during the subsequent 18 months. Multiple personality disorders, schizophrenia, and neurotic behaviors did not increase after the hurricane. The problems that occurred after the hurricane were usually problems of coping and communicating with families, friends, and support groups. The fact that many of the problems revolved around communication and a need for emotional expression lends evidence to the need for developing support groups to help victims process emotions and information after a disaster.

The UTMB has implemented a plan for developing support groups in the event that another hurricane strikes Galveston. Educational programs have been conducted for administrators, including identification of the stress reaction, proper supervisor's response to employees undergoing stress, needs of the victims, and how support groups can help. The response to these programs has been encouraging. Support groups are now included as part of the hurricane preparedness program. Department heads have agreed to identify a department liaison person to assist employees in using the social support group.

The support groups will be facilitated by a mental health professional. Their purpose is not to conduct group therapy, but to help with the group process, to help individuals feel more comfortable with expressing their feelings and emotions in a group setting, and to identify individuals who may be suffering from more extreme psychopathology. Departments will be encouraged to identify people who are candidates for a disaster support group, based on the extent of loss and the apparent need for social support. The recovery support groups will be limited to an eight-week period. They will be held during regular working hours and employees will be paid for the time that they participate in the recovery group. It is important to emphasize that these are not psychotherapy groups. The purpose of the group is to help employees work through their experience of a disaster, to help them develop effective problem-solving techniques, to provide them with information where available and appropriate, and to allow them an opportunity to express their thoughts and feelings about the disaster as a necessary step for coping with the aftermath. It is by providing an opportunity to do these things that we hope to minimize the potential for negative impact of a hurricane on both the physical and mental health of the employees.

There are at least two important features of our program that should be highlighted. The first is that the program is not just set up for disasters, but is part of an existing EAP and therefore will not be terminated at a certain point in time. By being part of an EAP, the hurricane recovery effort can begin immediately after the hurricane and can continue for as much as two years after the disaster has occurred. There are employees who are still in the process of recovering from Hurricane Alicia and who are not doing a particularly effective job of coping with the resulting problems. The second essential feature is that the program has been developed by an employer for both human and practical reasons. As an institution, the UTMB has a social responsibility to its employees. The practical aspect is that people are going to lose time from work after a hurricane. A recovery support group program can be an effective method to reduce lost time, accidents, and errors that will occur on the job. In community efforts to organize recovery programs, it is essential to include employers, particularly the large employers, in an area. Employer involvement in disaster recovery is in the best interest of the employees and the organization.

References

Bates, F.C. 1963. *The Social and Psychological Consequences of a Natural Disaster*. Washington: National Academy of Science.

Berren, M.R., A. Beigel, and G. Barker. 1982. "A Typology for the Classification of Disasters: Implications for Intervention." *Community Mental Health Journal* 18(2): 120-134.

Frederick, C. (ed.). 1978. *Training Manual for Human Service Workers in Major Disasters*. Rockville, MD: National Institute of Mental Health.

Hall, P.S., and P.W. Landreth. 1975. "Assessing Some Longterm Consequences of a Natural Disaster." *Mass Emergencies* 6: 55-61.

Kinston, W., and R. Rosser. 1974. "Disaster: Effects on Mental and Physical State." *Journal of Psychosomatic Research* 18: 437-456.

Kliman, A.S. 1976. "The Corning Flood Project: Psychological First Aid Following a National Disaster." *Emergency and Disaster Management*. Bowie, MD: Charles Press.

Logue, J.M., and H. Hansen. 1980. "A Case Control Study of Hypertensive Women in a Post Disaster Community: Wyoming Valley, Pennsylvania." *Journal of Human Stress*. June: 28-34.

Melick, M.E. 1978. "Life Change and Illness: Illness of Males in the Recovery Period of a Natural Disaster." *Journal of Health and Social Behavior* 19(3): 335-342.

Ollendick, D.G., and S.M. Hoffman. 1982. "Assessment of Psychological Reactions in Disaster Victims." *Journal of Community Psychology* 10: 157-167.

Quarentelli, E.L., and R.R. Dynes. 1977. "Response to Social Crisis and Disaster." *Annual Review of Sociology* 3: 23-49.

Titchner, J.L., and F.T. Kapp. 1976. "Family and Character Change at Buffalo Creek." *American Journal of Psychiatry* 133: 295-299.

LEGAL ISSUES

CONSTITUTIONAL ISSUES IN POST-HURRICANE RECONSTRUCTION PLANNING

Richard Hamann, Esq.

University of Florida Law Center

Introduction

In the aftermath of a destructive coastal storm, state and local government officials with land-use responsibility have an opportunity, through control of reconstruction, to avoid recreating the circumstances which led to disaster. A variety of regulatory measures are available. Building codes could be upgraded to require deeper foundations, stronger bracing and tiedowns, or higher elevations. Requiring the relocation of buildings further inland from the shore, through increased oceanfront setbacks, could allow for dissipation of storm forces and/or movement of the dune/beach system. If experience had proved that there were too many people to evacuate safely, an effort might be made to reduce the number of allowable dwelling units in the rebuilding effort. The reuse of certain parcels for construction purposes might be deemed so unsafe or so harmful to public resources that rebuilding of structures could be prohibited. To use an extreme example, the rebuilding of a house might not be allowed on a parcel permanently submerged offshore of the beach by an avulsive shoreline change.

Government might also choose to reduce its own exposure to future loss. Publicly owned structures and facilities could be redesigned, relocated, or abandoned to reduce the costs of repair and avoid probable future losses. If, for example, the expense of rebuilding a destroyed bridge is considered excessive, considering levels of use, available funds, and the probability of recurring damage, it could be relaced by a ferry service.

This paper will explore the potential for legal challenges to such action based on constitutional theories and will discuss the role of scientists and planners in the legal debate. The constitutional issues likely to be raised are ancient legal concerns—due process, equal protection, the "taking" clause—but they cannot be resolved by lawyers alone. The answers to those fundamental legal questions requires an interplay between scientific information and value judgments. "What is the law?" cannot be answered without knowing "What are the facts?"

Constitutional Theories

State and local authority over post-disaster reconstruction is based on the police power. The police power is the inherent, sovereign authority, reserved to the states by the Constitution, to promote the public health, safety, and welfare through regulation or other means. Balanced against this reservation of public power are various individual rights and liberties guaranteed by amendments to the Constitution. Of primary relevance are the rights of due process and equal protection and the prohibition against "taking" private property.

Due Process and Equal Protection

Due process and equal protection are guaranteed by the Fourteenth Amendment. Due process requires that an exercise of the police power be properly authorized by enabling statutes and procedurally fair. Of more substantive importance, due process requires that the exercise of power bear a reasonable relationship to the attainment of valid governmental objectives [1]. Equal protection analysis is similar, requiring that any classifications be rationally based [2].

The scope of permissible objectives has expanded greatly in recent years. The commonly accepted goals of coastal development regulation are uniformly upheld as valid police power objectives. Protecting the safety of those who might occupy coastal structures or be struck by their debris or who would be endangered in rescue missions are well-recognized, legitimate objectives [3]. Protecting the public health against the discharge of inadequately treated sewage is also recognized as valid [4]. Aesthetics [5], recreational access [6], water quality [7], fish and wildlife habitat [8], and the integrity of the beach/dune system [9] have all been treated as legitimate objectives for police power regulations in recent cases. Many of these objectives were not considered valid until relatively recently. As scientific and technical understanding have grown and been translated into popular understanding, the scope of the police power has grown to encompass new goals as legitimate and necessary.

There has been similar development in judicial support for the means chosen to attain the objectives of the police power. Due process requires that the means chosen for implementation be reasonably necessary to accomplish the purpose [10]. They must bear "a real and substantial relation to the object sought to be attained." [11]

Because the reasonableness of an ordinance can only be evaluated after considering a broad range of factors, the courts have not developed, and probably cannot articulate, specific criteria for defining the limits of substantive due process. Determination is made on a case-by-case basis by reference to the facts and circumstances of the situation. What is reasonable for one situation may well be unreasonable in another.

Considerations of public policy are important. The courts are, in essence, balancing two conflicting interests: the right of the individual to freely use property and the right of the public to protect broader social and environmental interests. To determine whether an ordinance is reasonable, the courts must weigh the strength of public interest factors supporting the regulation. A court's perception of the public interest will depend on its understanding of the facts of the case and of how the development at issue fits into larger patterns of land and water use.

A growing sensitivity of the courts to the need for stringent coastal management is evident in recent decisions. The use of setbacks [12], height limitations [13], development moratoria [14], transferable development rights [15], impact fees [16], mandatory dedications [17], floodplain zoning [18], wetland regulations [19], and stormwater runoff controls [20] have all been upheld as reasonable exercises of the police power.

A decision by Florida's Fifth District Court of Appeal, upholding the reasonableness of a beach setback regulation, is illustrative. In *Indialantic v. McNulty* [21], the property in question was located between a road and the Atlantic Ocean. The town had established a regulatory dune line running approximately 20 feet seaward of the road and required residences to be set back 25 feet landward of the line. Building was thus effectively prohibited. The owner sought a variance to construct a house on stilts over the dunes; the variance was denied.

The opinion is exceptionally well reasoned. In applying the applicable legal principles, the court considered the important natural resource functions of the dune system. In addition, the dune area is a hazardous location for building. As the court stated:

> There can no longer be any question that the 'police power' may be exercised to protect and preserve the environment...the wetlands and coastal areas are places of critical concern because of their important role in protecting the inland regions against flooding and storm danger. The ordinance in this case passes constitutional muster because it was not shown to be in any way arbitrary or discriminatory, or more severe or strict than necessary to achieve a valid police power purpose. [22]

Where the reasonableness of an approach has not been well substantiated, however, particularly where substantial interference with use of property resulted, the courts have been willing to invalidate regulations.

In *City of Boca Raton v. Boca Villas Corp.* [23], the citizenry established by initiative and referendum a 40,000-unit development cap for the entire city. A property owner sued, claiming there was no rational relationship between the density cap and a permissible municipal purpose. The court agreed, noting that the city planning department had never been consulted on the need for the cap. The court also found it significant that the director of the planning department testified that other than "community choice" he knew of no compelling reason for imposing a permanent, fixed limitation on population or dwelling units [24]. Finally, the court considered the relationship of the cap to such aspects of community welfare as utility services, schools, fiscal soundness, water resources, air quality, noise levels, and comprehensive planning. In each case, the court found the evidence inadequate to show that the cap promoted that aspect of community welfare [25].

In a more recent case, however, another coastal community was able to prove a rational relationship between a density cap and the health, safety, and welfare of the people. In *City of Hollywood v. Hollywood, Inc.* [26], the Fourth District Court of Appeal again decided the validity of a cap. In that case, the City of Hollywood had placed a 3,000-unit cap on an area adjacent to the Atlantic Ocean that was held reasonable because it was based on rational consideration and study [27].

> A multitude of factors were taken into consideration over and above traffic. Water and sewer capacities were measured as was the provision of services such as fire and police protection. The question of how to evacuate the residents in a hurricane with only two possible escape routes to the mainland, one at each end, was also considered concomitantly with maintenance of the dune line to protect against storm ravage. The fact that this, as the developer admitted, is the last undeveloped beach area on the Gold Coast, is filled with desirable rare flora, is ecologically sensitive and crying out for environmental protection, is in desperate need of open space and easy public access to the ocean, were all addressed and considered in agonizing detail. . . .the question of shadow on the public beach from these proposed monolithic structures was also addressed. . . .last but not least, the record shows that much thought was given to aesthetics. . . .the record is replete with comprehensive plans, studies, reports, public meetings, and actual discussions with the developer over a period of years. Unlike the Boca Raton case, the City of Hollywood did not present its community purpose in the abstract, but presented a more than adequate case for the proposition that the proposed cap would contribute substantially to the public health, morals, safety and welfare of its citizens.

The quality of study and of documentation was an important factor distinguishing the two cases. There is little doubt that police power action to prevent recurrence of a disaster will receive sympathetic judicial review. But significant levels of interference with future use receive stricter scrutiny in the balancing process and their reasonableness should be supported by thorough, detailed study.

The "Taking" Clause

The Fifth Amendment provides ". . . nor shall private property be taken for public use, without just compensation." Although intended to prevent uncompensated seizures of title or of physical possession, the Supreme Court has interpreted the clause to prohibit police power actions that have an equivalent effect [29]. As with the determination of reasonableness, however, no clear test has emerged from hundreds of "takings" cases as to when regulation or other police power action takes private property. Determination depends upon the specific facts and circumstances of each particular case. Most opinions reflect a balancing of the factors and the use of several tests.

The effect of regulation on the monetary value of property is a major factor in the analysis of "takings" claims. Decisions invalidating land-use regulations as takings have done so primarily because property values were too severely depressed. The law is clear, however, that diminution in the value of regulated property does not, in and of itself, establish a "taking." [30] It is only one factor to be considered by the courts. Numerous decisions have sustained stringent regulations against such challenges even though property values were very substantially reduced [31].

From another perspective, the courts may consider whether sufficient value or use remains for the parcel. A "taking" occurs only when there has been a virtual destruction of the property interest. If the potential exists for the owner to make some reasonable use of the regulated property, then no "taking" will be found [32]. Recent versions of this test require consideration of whether the government action results in a less than economically viable use of the property [33] or whether it forecloses all reasonable investment-backed expectations [34].

The courts have incorporated several important qualifications into "takings" theory. An entire parcel is examined, not just the portion subject to the greatest restrictions [35]. In examining land subject to oceanfront setbacks, for example, the remaining value and usefulness of the entire lot is considered. Assuming valuable economic uses are left to the remainder, a "taking" is unlikely to be found.

But what about the situation where application of the necessary setbacks results in a lot with no remaining buildable portions? The owner is likely to argue a loss of reasonable use (or of all reasonable investment-backed expectations). An appropriate response is that it is not reasonable to build in a dangerous location [36]. Further, one should have no legitimate expectation of being able to rebuild or otherwise use property if to do so would threaten harm to the property of others or to public resources [37]. Denial of an unreasonable use is not a "taking."

In *Spiegle v. Borough of Beach Haven* [38], the application of a dune protection setback to several lots was challenged as a "taking." There was a prohibition against building residences seaward of the setback line. The court looked at the circumstances of each lot individually. On one lot, the court found that it was feasible to build a residence and held that it was a "taking" to deny that use. As for the other lots, there was no "taking" [39], because:

> . . . although it may be possible from a strictly engineering standpoint to erect a residential structure on the site with supporting utilities, it would not be safe or economically feasible to do so from a common-sense standpoint. As with any health or safety regulation, the interest of one property owner must be subjected to some degree to the welfare of the general public.

A similar type of restriction, prohibiting building in the floodway of the Klamath River, was upheld in *Turner v. County of Del Norte* [40]. There was evidence of previous flooding and evidence that if buildings were constructed they would likely be destroyed, endangering their occupants and the owners of other property. Rejecting a claim of "taking," the court noted that the ordinance "imposes no restrictions more stringent than the existing danger demands" [40].

The "taking" clause was enacted to prevent government from acquiring unjustified benefits. Many cases have held that it does not prohibit the government from preventing harm [42]. The distinction between prevention of harm and acquisition of benefits is important in "taking" cases. There is a long tradition of using the police power to prevent nuisance-like or otherwise harmful effects. The abatement of hazards to life and property, unsanitary conditions, beach erosion, or impediments to the public use of public beaches would arguably fall within the scope of permissible regulation despite the loss of expected profits by an affected landowner.

Planning to Prevail on Constitutional Issues

The balancing of public and private interests inherent in constitutional analysis of police power action is necessarily somewhat subjective. The results of litigation are therefore difficult to predict. Advocates of post-hurricane regulation, however, enjoy tremendous advantages in the legal debate. The problem being addressed is not speculative and abstract; it will have happened. The causes and results are likely to be clearly evident. If buildings failed because of inadequate foundations, the reasonableness of strengthened building codes is not likely to be seriously questioned. The courts will not seek to second-guess an expert determination that setbacks should be established 300 feet from the water rather than 100 feet.

Nevertheless, severe restrictions on future land use, and particularly prohibitions on rebuilding, may be scrutinized. The probability of a successful defense can be enhanced if several considerations are taken into account in drafting and implementing post-hurricane reconstruction plans. The planning process should anticipate and prepare to counter legal challenges by incorporating the advice which follows:

1. Leave open the possibility of development. Rather than prohibiting reconstruction, performance standards and restrictive criteria might be imposed that are sufficient to protect the public interest. The denial of an application for one proposal does not foreclose the possibility of a future approval made possible through changes in design or improvements in technology.

2. Offer the owner incentives for compliance with the plan. An opportunity to transfer or sell density credits may be an appropriate incentive or may serve as mitigation of financial impacts for the owner. If purchase is considered, beware of saying or writing anything that can be used as evidence of an underlying intent to reduce the costs of acquiring the property for use as a public park.

3. Base the plan on the very best available technical information. Plan comprehensively, relating individual cases to larger issues and objectives. Here the role of the scientist or planner is most critical. Not only must the best possible plan be developed, but it must be as well substantiated as possible.

4. The magnitude of problems addressed by the plan should be well documented. Preparations should be made in advance to gather information on the effects of the storm that triggers implementation of a reconstruction plan. The consider-

ation of less restrictive alternatives should also be documented. Remember that someday the documents may be introduced into evidence in support of the plan.

5. Summarize the data and conclusions in clear, concise language that any layperson can understand. Use photographs, drawings, maps and other illustrations. The plan should be popularized and communicated to the general public to the greatest extent possible. A plan that is widely understood and supported is likely to be upheld. Good facts make good law.

Notes

1. *Mugler v. Kansas*, 123 U.S. 623 (1887).
2. See, e.g., *Responsible Citizens v. City of Asheville*, 302 S.E. 2d 204 (N.C. 1983).
3. See, e.g., *Turnpike Realty Co. v. Town of Dedham*, 284 N.E. 2d 891 (Mass. 1972).
4. See, e.g., *Milardo v. Coastal Resources Management Council of Rhode Island*, 434 A. 2d 266 (R.I. 1981).
5. See, e.g., *City of Hollywood v. Hollywood, Inc.*, 432 So. 2d 1332 (Fla. 4th DCA 1983).
6. See, e.g., *Sea Ranch Ass'n v. California Coastal Comm'n*, 527 F. Supp. 390, 395 (N.D. Cal. 1981).
7. See, e.g., *City of Annapolis v. Annapolis Waterfront Co.*, 396 A. 2d 1080 (Md. App. 1979)
8. See, e.g., *Potomac Sand and Gravel Co. v. Governor of Maryland*, 266 Md. 358, 293 A. 2d 241 (1972).
9. See, e.g., *Indialantic v. McNulty*, 400 So. 2d 1227 (Fla. 5th DCA 1981).
10. *Goldblatt v. Town of Hempstead*, 396 U.S. 590 594 (1962).
11. *Nebbia v. New York*, 291 U.S. 502, 524 (1934).
12. See, e.g., *Indialantic v. McNulty*, 400 So. 2d 1227 (Fla. 5th DCA 1981).
13. See, e.g., *City of Hollywood v. Hollywood, Inc.*, 432 So. 2d 1332 (Fla. 4th DCA 1983).
14. See, e.g., *Franklin County v. Leisure Properties, Ltd.*, 430 So. 2d 475 (Fla. 1st DCA 1983).
15. See, e.g., *Penn Central Transportation Co. v. New York City*, 438 U.S. 104 (1978).
16. See, e.g., *Contractors and Builders Association of Pinellas County v. City of Dunedin*, 329 So. 2d 314 (Fla. 1976).
17. See, e.g., *Hollywood, Inc., v. Broward County*, 431 So. 2d 606 (Fla. 4th DCA 1983).
18. See, e.g., *Young Plumbing and Heating Co. v. Iowa Natural Resources Council*, 276 N.W. 2d 377 (Iowa 1979).
19. See, e.g., *Sibson v. State*, 336 A. 2d 239 (N.H. 1975).
20. See, e.g., *City of Annapolis v. Annapolis Waterfront Co.*, 396 A. 2d 1080, 1089 (Md. App. 1979).

21. 400 So. 2d 1227 (Fla. 5th DCA 1981).
22. *Id.* at 1232.
23. 371 So. 2d 154 (Fla. 4th DCA 1979).
24. *Id.* at 155.
25. *Id.* at 156-157.
26. 432 So. 2d 1332 (Fla. 4th DCA 1983).
27. *Id.* at 1334-36.
28. F. Bosselman, D. Callies, and J. Banta, 1973. *The Taking Issue.* Washington: U.S. Government Printing Office.
29. *Pennsylvania Coal Co. v. Mahon,* 260 U.S. 393, 43 S. Ct. 158, 67 L. Ed. 322 (1922).
30. *Penn Central Transportaion Co. v. New York City,* 438 U.S. 104, 131 (1978).
31. In *Euclid v. Ambler Realty Co.,* 272 U.S. 365 (1926), zoning was upheld despite a 75% diminution of value. The owner of the regulated brickworks in *Hadacheck v. Sebastian,* 239 U.S. 394 (1915), sustained a 93% loss in the value of his property (from $800,000 to $60,000), yet the regulation was sustained. The practical effect of *Goldblatt v. Town of Hampstead,* 369 U.S. 590 (1962), was to prohibit further operation of a rock and gravel mine. Finally, in *Miller v. Schoene,* 276 U.S. 272 (1928), the regulated property, cedar trees, was completely destroyed.
32. *Moskow v. Commissioner of the Dept. of Environmental Management,* 427 N.E. 2d 750 (Mass. 1981).
33. *Agins v. City of Tiburon,* 447 U.S. 255, 100 S. Ct. 2138, 65 L. Ed. 2d 106 (1980).
34. *Penn Central Transportation Co. v. New York City,* 438 U.S. 104 (1978).
35. *Graham v. Estuary Properties, Inc.,* 399 So. 2d 1374 (Fla. 1981).
36. *Spiegle v. Borough of Beach Haven,* 281 A. 2d 377 (N.J. App. 1971); *Turner v. County of Del Norte,* 24 Cal. App. 2d 311, 101 Cal. Rptr. 93 (Cal. App. 1972).
37. *Just v. Marinette County,* 56 Wis. 2d 7, 201 N.W. 2d 761 (1972).
38. 281 A. 2d 377 (N.J. App. 1971).
39. *Id.* at 387.
40. 24 Cal. App. 2d 311, 101 Cal. Rptr. 93 (Cal. App. 1972).
41. *Id.*
42. See, e.g., *Graham v. Estuary Properties, Inc.,* 399 So. 2d 1374 (Fla. 1981).

THE NFIP AND DEVELOPED COASTAL BARRIERS

Alexandra D. Dawson, Esq.

Hadley, Massachusetts

Introduction: Coastal High Hazard Areas

The National Flood Insurance Program (NFIP) is administered by the Federal Emergency Management Agency (FEMA) under the National Flood Insurance Act of 1968, as amended (42 U.S.C. secs 4001 et seq.). Since this paper was prepared for a conference on the management of developed coastal barriers, discussion will largely address "coastal high hazard areas" or "V-zones." These areas are described in NFIP regulations as ". . .subject to high velocity waters, including but not limited to hurricane wave wash or tsunamis" (44 C.F.R. sec. 59.1). These include the ocean-facing shorelines of nearly all coastal barriers.

The landward extent of the V-zone is not defined in the regulations, but for purposes of mapping is considered to include areas within reach of a three-foot breaking wave during a 100-year storm surge (fig. 23.1). While this paper is chiefly concerned with the ocean-facing shoreline, it should not be forgotten that coastal barriers also have a bayside shoreline which is normally subject to storm surge but not breaking waves. Bayside flood hazard areas are usually designated as "A-zones" on NFIP maps. (For a general discussion of the NFIP in relation to coastal management, see: Platt, 1978 and 1985.)

Coastal high hazard zones, by definition, are exposed to intense storm activity, to erosion, and to wind damage. Despite these perils, the NFIP does provide flood insurance to both existing and new structures in V-zones, except where prohibited by the Coastal Barrier Resources Act of 1982. Coastal communities (including counties) must, however, comply with FEMA's land-use and building standards for new construction or substantial improvements in coastal high hazard areas. These requirements, as amended in the *Federal Register* of September 4, 1985, appear as an appendix to this paper. In essence, they provide that:

1. All new construction must be located landward of mean high tide
2. The lowest floor must be elevated above "base flood level" (100-year flood elevation)

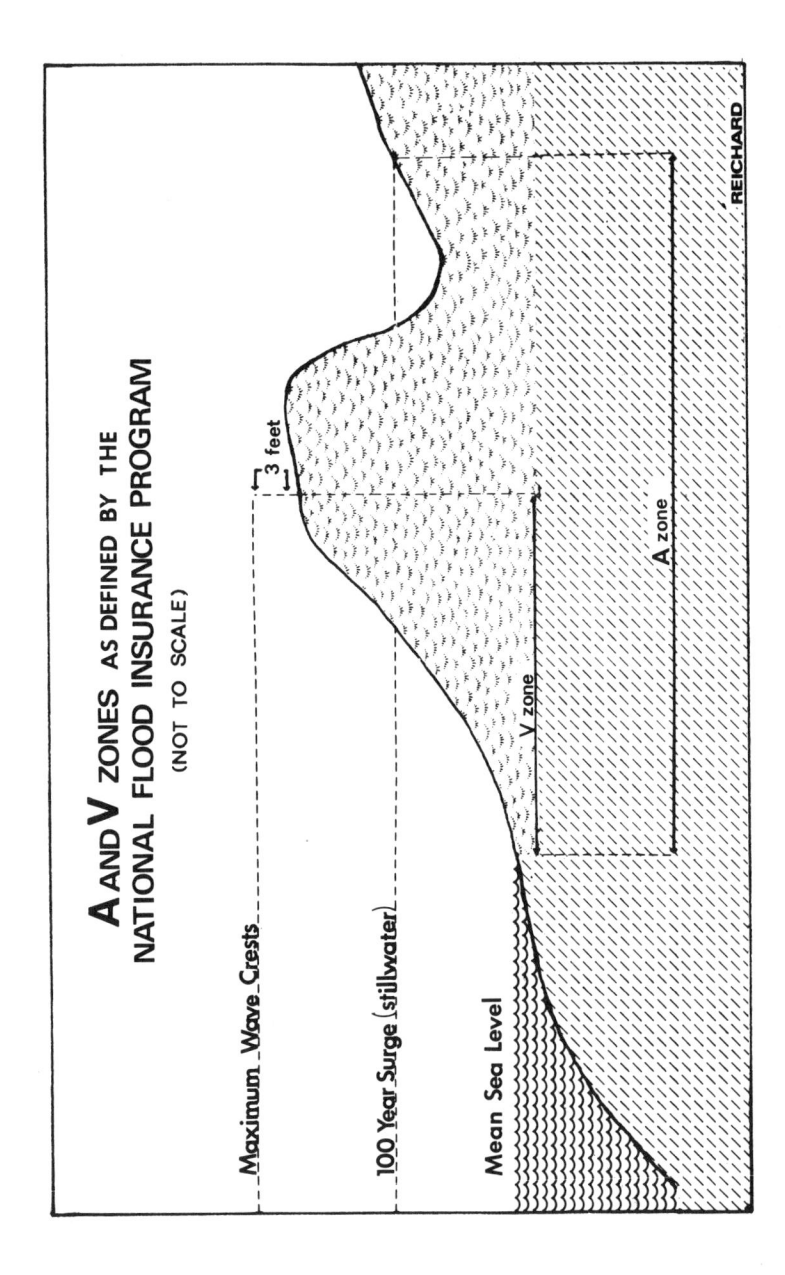

Figure 23.1. Cross-section of Coastal 100-Year Floodplain (A Zone) and Coastal High Hazard Zone (V Zone) as defined by the National Flood Insurance Program

3. Structures and pilings must be anchored so as to prevent flotation, collapse, and lateral movement
4. The space between grade level and the lowest floor may only be enclosed by "breakaway walls" or other means which do not obstruct wave action
5. The use of fill is prohibited for support of structures
6. Mobile homes are prohibited in V-zones (except in existing mobile home parks or subdivisions)
7. Human alteration of sand dunes and mangrove swamps is prohibited in V-zones.

Data Base: What Do We Know About Developed Coastal Barriers and the NFIP?

Developed coastal barriers account for approximately 280,000 acres or 17% of the total coastal barrier area of 1,645,000 acres. (Of the rest, 885,000 acres are owned by public or private conservation organizations [Sheaffer and Roland, 1981].) Almost 40% of the total coastal barrier beach mileage is developed: 1,048 of 2,685 miles. Protected mileage is slightly less; 26% is undeveloped and unprotected. Developed beach mileage varies widely from state to state, from zero in Mississippi to 100% in New Hampshire (U.S. DOI, 1985). In 1981, the Federal Insurance Administration anticipated that 3,000 new structures would be built annually in V-zones. Ninety communities are totally within developed coastal barriers. From 1978 to 1981, 64,000 claims totalling $43 million were paid in these communities (FEMA, unpublished data).

In 1982, FEMA estimated premium coverage under the NFIP at $10-15 billion for developed coastal barriers, comprising 200,000 policies. All estimates relating to this coverage are, however, subject to question because FEMA does not distinguish insurance on barriers from insurance in the rest of coastal communities. Allocation was accomplished by visual estimate from studying maps and prorating coverage (by "best guess") between coastal barrier and mainland policies in each community.

Coastal communities generally, as of November 1984, contained 70% of all NFIP policies, 73% of coverage, but only 54% of claims and claim dollars. In past years, per-capita damages have been twice those of inland claims for buildings, three times for building contents (Miller, undated (a) & (b)). Insurance premiums in coastal areas have been twice as costly as inland premiums (ASFPM, 1982). (However, these figures may have been "skewed" by Hurricane Frederic in 1979.) The figures cover all flooding damage in coastal communities, not just ocean storm damage in V-zones. FEMA staff believe that the low figure for claims versus premiums is partly attributable to frequent low-level flooding of "finished" basements in inland houses. In 1983, FEMA reduced insurance coverage for the contents of such basements, although it will still insure some appliances and structures. Another reason given for the magnitude of the figures is that coastal storms are less frequent than inland floods but cause more damage when they occur. In support of this, Hurricane Alicia, a medium-sized storm, generated $100 million in NFIP claims (National Research Council, 1984, p. 139).

H.C. Miller believes that 300 communities produce 90% of the claims (personal communication). FEMA states that in 1983, the top 150 claimant communities had 55% of the total claims and 70% of the claim dollars. A survey in Massachusetts disclosed that 29 developed barrier communities account for 51% of the NFIP coverage and 53% of the policies in the entire state. Only three of the 29 are not in the program. One of these is protected by a barrier island; another is very small; the third is developed and exposed (Mass. DWR, 1983). Less than one-third of the Massachusetts coastal communities are still in the emergency phase of the program.

As of 1984, 81% (1130) of all Atlantic and Gulf Coast communities were qualified for the regular phase of the NFIP. Of the 19% still in the emergency program, one-half are in New England (100 in Maine, 23 in Massachusetts, 10 in New Hampshire). Others are in New York, Virginia, and southern North Carolina. Delays in New England may also reflect difficulties in computing "wave-run-up" models in areas with steeper offshore slopes. Also, these numbers may in part reflect the size of communities rather than the importance of those still in emergency status. FEMA does not distinguish in its numbers between a small New England town and a large county elsewhere; both count as "communities."

The 81% of communities in the regular program account for 97% of the coastal flood insurance policies and 98% of the coastal dollar coverage. The regular/emergency ratio for the nation as a whole is about 60/40, up from 50/50 in 1982 (FEMA, 1983). The overall participation rate in the program is over 17,000 out of 21,000 flood-prone communities (Gibson, 1984, p. 156).

What lies ahead for developed coastal barriers? The U.S. Department of the Interior (1985) predicts that much more construction will occur during the coming years in the now partially developed area. Coastal areas are developing much faster than the rest of the country (Sheaffer and Roland, 1981; Platt, 1978). Official estimates show that population increases from 1945 to 1975 ranged from a low of 39% in the New York Bight and 47% in New England to 99% in the Mid-Atlantic, 110% in Texas, 201% in the Georgia Sea Islands, 208% in Atlantic Florida, 302% in the eastern Gulf, and 309% in Louisiana; the median growth rate was 153% (U.S. DOI, 1985). The lower figures for the Northeast may simply reflect the larger amount of development dating from earlier periods. It does seem likely, though, that growth since 1975 would reflect concentration in southern areas.

Although there are 130,000 annual housing starts overall in coastal communities (FEMA, 1980), as many as 90% of the insured buildings in V-zones are estimated by FEMA to predate the NFIP and thus are not in conformity with the program's standards for new construction (elevation, floodproofing, and so on.)

There is no reason to believe that flooding experience will limit future development: the Wharton School studies (Kunreuther, 1978) show that the longer people live in a coastal area, the less they fear storms—those who have experienced a major storm are only 21% less likely to rebuild. Coastal inhabitants tend to be fatalistic and prepare less for disaster than inland residents (Miller, undated (a)). Massachusetts has shown that a large majority of coastal residents prefer to rely on structural defenses paid for by federal funds and oppose relocation (Mass. DWR, 1983).

A study by the U.S. General Accounting Office (1982) found the existence of flood insurance to be a "marginal additional inducement" for development in coastal barrier communities. That is, people are determined to buy, sell, build, and live in coastal areas, regardless of the associated risks. Flood insurance availability is more influential in the few communities (such as in Rhode Island and Galveston, Texas) where banks previously denied mortgages for floodplain development. Other federal subsidies such as those for infrastructure development are more influential overall.

Mapping Issues in the National Flood Insurance Program

Practically all ocean-facing shorelines are mapped as V-zones. However, much of the coastal barrier area not facing the ocean, including bayside shorelines, is mapped as zones A or B (representing the 100- and 500-year floodplains, respectively). There is no information from FEMA as to whether all areas that should be designated as V-zones have been so mapped.

Many policies are misrated. A 1982 sample of 94 policies in five towns showed 34 incorrect ratings, 24 of them under-ratings. That is, the dwellings were rated as though in an A-zone, although they were in fact located in a V-zone. It was estimated that this underrating could be costing the federal government $25 million per year in unpaid premiums (U.S. GAO, 1982).

FEMA has no clear policy on remapping after a storm, although erosion may have shifted the boundaries of coastal beaches. Sometimes new maps are issued with boundaries changed so that areas formerly in the A-zone are included in a V-zone. If new maps are issued, buildings that were located in the A-zone on the old map are not re-rated when the boundaries are changed (FEMA, 1984). However, such a dwelling must be elevated to designated flood levels if "substantially" improved.

A major recent development in the NFIP is the addition of "wave heights" to the elevation standards in coastal V-zones. Previously, such standards were based on still-water elevations, which did not reflect potential damage from waves breaking above that elevation. Of the 636 communities with designated V-zones along the Atlantic and Gulf coasts, by January 1985, 418 had received new flood insurance rate maps (FIRMs) reflecting potential wave heights. Maps for the remaining coastal communities were to be completed by 1986.

There is always a delay between publication of such new requirements and their adoption by communities. FEMA does not operate by fiat and must give participating communities an opportunity to study, dispute, and vote upon any changes in land-use maps and regulations. In the interim, FEMA regulations were amended in 1981 (44 CFR 59.11 (e) (1-3)) to require that all new and "substantially improved" construction in V-zones be rated using available wave data with a minimum extra rise of 2.1 feet. Studies in FEMA Region I (New England) estimate that wave addition raises the location of the lowest structural member of the building to an average height of 15 to 17 feet above National Geodetic Vertical Datum of 1929 (NGVD) in steep beach areas, with the new elevations along a flat beach running about 13 to 15 feet above NGVD. This would be a rise of 2 to 5 feet above present requirements. Extreme wave heights can be as high as 25 feet above NGVD. The Washington office of FEMA reports that base-flood elevations over 20 feet are found only in Louisiana and northwest Florida; most are 11 to 18 feet elsewhere along the coast, and drop rapidly as one proceeds inland.

Erosion Issues

Coastal barriers are highly susceptible to erosion; development may increase local erosion rates dramatically. Windblown sand that formerly moved inland and then blew back onto the beach no longer travels freely, buildings create tunnels for wind erosion, and dunes are disturbed or destroyed by building and by pedestrian or vehicular traffic (IEP, Inc., 1983).

The NFIP defines erosion (44 CFR 59.1) as: "The process of the gradual wearing away of land masses." The NFIP does not insure against gradual, ordinary erosion. It does, however, cover unanticipated sudden erosion damages resulting from flooding. The definition of flooding in 44 CFR 59.1 includes:

> [T]he collapse or subsidence of land along the shore of a lake or other body of water as a result of erosion or undermining caused by waves or currents of water exceeding anticipated cyclical levels or suddenly caused by an unusually high water level in a natural body of water, accompanied by a severe storm, or by an unanticipated force of nature, such as flash flood or an abnormal tidal surge, or by some similarly unusual and unforeseeable event which results in flooding. . . .

At our current level of knowledge about "anticipated" coastal events, this definition is an invitation to controversy (FIA, 1977). Even if erosion is ongoing and not covered as such, damage may finally occur to a structure as part of a flood event and thus be covered in that way.

At its 1977 erosion conference, the Federal Insurance Administration (FEMA's predecessor) expressed interest in promulgating regulations that would establish a 30-year dune erosion line in coastal V-zones, seaward of which nothing could be insured (FIA, 1977). However, no such regulations were developed because it was difficult to find erosion data that would stand up to debate. FEMA does encourage communities to establish a minimum setback for anticipated erosion of at least a period of 30 years, the average length of a building mortgage.

It would be preferable to use a period of 50 to 70 years, the life of the average building (Ralph M. Field Associates, 1983).

Several states have established their own erosion-setback lines. Michigan has established a 30-year dune line for its coastal bluffs under state law (ASFPM, 1983). Georgia has established a "primary dune area" which extends 40 feet landward from the most seaward stable dune, within which only fences and boardwalks are allowed. The North Carolina administrative code establishes an "ocean erodible area" extending 30 times the annual erosion rate or a minimum of 60 feet landward (ASFPM, 1982). South Kingston, Rhode Island has attempted to limit use of "high flood danger" segments of its beach area to preclude dwellings (Ralph M. Field Associates, 1983). This last example illustrates some of the legal difficulties of regulation: the Rhode Island Supreme Court invalidated this bylaw as a "taking without just compensation" of a landowner's property (*Annicelli v. Town of S. Kingston*, 463 A.2d 133, 1983). This is not a typical decision, since courts have been upholding no-build regulation of flood-prone areas for over a decade. However, the bylaw had two particular problems: It emphasized preservation of open space and beaches over flood hazard mitigation, and it was applied in an area where 30 homes already existed. The reasoning of the court on the latter point shows how difficult it is to halt further building in developed areas.

In response to the erosion problem, FEMA also advocates beach nourishment and dune revegetation. But it advises against the placement of groins and nourishment of beaches by offshore excavation (Ralph M. Field Associates, 1983).

The first effect of erosion on insurability is the requirement that a community in the regular program "provide that all new construction within V-zones 1-30 on the community's Flood Insurance Rate Map is located landward of the reach of mean high tide" (44 CFR 60.3 (e) (3)). These requirements would not in themselves ban use of insurance proceeds to rebuild if the building is not to be "substantially improved" (as discussed below). Some states have tried to ban rebuilding under these circumstances, but found it politically difficult because of high land values and post-storm public sympathies (Platt and McMullen, 1980). FEMA regulations that appear to require communities to prepare erosion plans involving setbacks (44 CFR 60.5) have never been applied; they were intended only to apply to certain Great Lakes erosion areas that were never designated by FEMA.

In all states except Massachusetts and Maine, coastal land between mean high and low water is owned by the state, with the abutters having only limited rights. Where erosion causes buildings to impinge into this area (or more accurately, causes the reverse), there may be a de facto case of trespass. Texas has in fact taken title to properties located seaward of the new vegetation line after Hurricane Alicia (National Research Council, 1984; see McCloy and Huffman paper).

Periodic revegetation of naturally occurring dunes is a more cost-effective way of preventing erosion than the erection of groins and other structural defenses (Ralph M. Field Associates, Inc., 1983; IEP, Inc., 1983). In 1976, FEMA regulations were amended to require that a community in the regular program must "prohibit man-made alteration of dunes and mangrove stands within V-zones 1-30 on the community's rate map which would increase potential flood damage" (44

CFR 60.3 (e) (8)). The wording was intended to limit the federal role to flood damage reduction, as opposed to environmental protection. It is certainly vague enough to allow a community to let a builder relocate the natural dune, or at least to construct buildings right over a dune from the back side (Platt, 1978). This kind of construction and use disturbs dunes, and mining sand offshore to create new dunes increases erosion. In Ocean City, Maryland, 12-foot dunes built by the Army Corps of Engineers after a 1962 storm were either washed away or removed within two decades (IEP, Inc., 1983).

Regulations adopted by Massachusetts in 1983 under its Wetlands Protection Act (Mass. G.L.A., ch. 131, sec. 40) provide performance standards for buildings on or near dunes which tend to protect the dunes, although they do not establish an erosion-setback line. The law is, however, somewhat ambiguous as it applies to repairs and replacement of existing buildings. Although it makes no exemption for such work if it involves new encroachment on protected resource areas, many regulators and at least one court have assumed such an exemption, perhaps by analogy to standard principles of zoning law.

States and communities cannot be relied upon to protect dunes on a local basis. The federal regulation should be amended to forbid any alteration of dunes except for restoration and revegetation. Most community officials say they would accept tougher standards if FEMA promulgated them (Miller, undated (b)).

Enforcement

Regulation of "Substantial Improvements"

If buildings are to be repaired or reconstructed at a cost of more than 50% of their pre-flood market value (excluding contents), FEMA requires local communities to apply its standards for new construction to mitigate future damage. These involve elevation, floodproofing, more expensive construction, tie-downs for trailers, and so on. The determination as to the percentage of value involved is left to the local building inspector or equivalent official, who bases his calculations on tax assessment (adjusted to true market value) versus repair estimates submitted in the application for a building permit. There are several problems with this system. Because most state and/or local building codes do not require the applicant for a building permit to include all improvements, submitted repair figures tend to be lower than actual costs (Miller, 1979). Also, since the regulation defines substantial improvement in terms of repair cost, the damage cost, which is a more germane figure, cannot be used by FEMA. Thus an owner who defers certain repairs may avoid compliance with applicable mitigation regulations although the structure was more than 50% destroyed.

Opinions differ as to the extent to which nonenforcement of the "substantial improvement" provision is a problem. Experience in New England after the 1978 blizzard and in Alabama after Hurricane Frederic indicates that the requirement can be implemented after large coastal storms. On the other hand, it is conceded that local officials may tend to call repairs nonsubstantial in marginal situations, involving perhaps up to 70% of pre-flood values. In past years, implementation was said to be "virtually nonexistent," but FEMA states that it no longer downplays the issue. The agency has repudiated its own 1980 study that stated there was no reason for concern. The study estimated that of the nation's 72.6 million dwellings, only 126,000 (0.2%) are "substantially improved" annually (by the FEMA definition of the term) for any reason whatsoever: Of these only 18,200 are located in flood-prone areas, in comparison with 130,000 new housing starts taking place in those areas every year (FEMA, 1980). The reasoning ignored the fact that nonconforming structures will continue to be more subject to damage and more highly subsidized than new or "substantially improved" structures, which are supposed to pay their own way (through actuarial rates) in the NFIP. Thus structures that do not come up to new construction standards are recurrently as well as artificially subsidized. A change in the regulation and a study of enforcement are required.

Breakaway Walls Below Elevated Structures in V-Zones

A limited study by the U.S. GAO (1982) confirms remarks found throughout the literature that there is spotty enforcement of the requirement that space below the lowest floor of elevated structures in V-zones must be kept "free of obstruction or be constructed with 'breakaway' walls [and] not be used for human habitation." In 1980, FEMA actually tried to promulgate a regulation forbidding any solid walls whatever under these elevated structures. The Office of Management and Budget objected to this as "federal intrusion into local building affairs." The GAO, however, continues to advocate such a role (GAO, 1982). Some solid walls are built with permission; others are altered without a building permit. FEMA now charges substantially higher premiums for any solid walls under the elevated area and may also refuse to pay claimants if they have violated this or other floodplain management regulations (GAO, 1982).

Aside from the question of whether the premium increase is adequate, there are other reasons to enforce the requirement for breakaway walls. Non-breakaway walls may imperil not only the structure they are under, but the surrounding area as well. They offer an obstacle to storm waves that can change flows and increase scouring, undermining building supports. Studies in Galveston after Hurricane Alicia showed that scouring can cause unexpected undermining of even properly designed buildings (Miller, 1983). Communities or states may now require open or latticed underspaces, but such a requirement is politically difficult to impose on a community-by-community basis.

General Monitoring of the NFIP

FEMA staff and others disagree as to whether the agency has the resources to do an adequate enforcement job for the NFIP. Originally FEMA had proposed to review one-fifth of all communities in the regular program every year, including meeting with local officals. It has not met this goal (GAO, 1982) and now believes it can achieve enforcement by concentrating on targeted communities. Pursuant to a 1981 regulatory change, FEMA does not assess individually each new building in V-zones but relies on certification of elevation by a registered engineer, architect, surveyor, or community official on its Post-Construction Elevation Certificate, which is filled out by every applicant for insurance in V-zones (44 CFR 60.3 (e) (4)). In addition, FEMA concentrates on applying actuarial rates to construction that is not built or substantially improved according to its standards, rather than to deny insurance or to suspend communities for nonenforcement. The Washington office of FEMA believes that the raising of rates since 1980 has become an effective tool, whereas suspension would penalize the whole community. Suspension is also costly and unpopular and should be used as a last resort. Actuarial rates are generally viewed as a fair tactic.

FEMA has sued Jefferson and St. Bernard Parishes in Louisiana for failing to enforce the building standards they had adopted when they joined the NFIP, and also for failing to maintain drainage facilities in those flood-prone areas. The suit is still in trial (*U.S. v. Parish of St. Bernard*, 5th Circuit, U.S. Ct. Appeals #83-3557). Issues so far revolve around the applicability of federal common law "nuisance" decisions and whether the U.S. government can collect from the communities $100 million in claims paid out for storms in 1978 and 1980, or whether its sole remedy under the law is to suspend the communities if they refuse to conform.

FEMA has moved from selling insurance to letting private insurance companies sell and service policies. The effects of the new "write your own" insurance program are as yet unknown. FEMA staff do not think that local insurance agents will be motivated by federal subsidies to collude with illegal behavior.

Relation to Disaster Relief Program

One major purpose of the NFIP was to lower taxpayers' costs for disaster relief by shifting a significant portion of the cost to "users" of flood-prone areas (League of Women Voters, 1982). It is not clear whether disaster relief costs have actually decreased in coastal areas, partly because of the occasional nature of coastal storms and partly because of the huge increase in development there as well as the high costs of construction and repairs.

A major form of hidden tax subsidy not generally thought of as "disaster relief" is the continuation of provisions in the Internal Revenue Code allowing taxpayers to deduct casualty losses over 10% of their adjusted gross income. This is a major form of subsidy to coastal barrier development where losses may be great and taxpayers wealthy (DOI, 1985). [This may be changed by tax reform legislation in 1986, ed.]

A principal inducement to join the NFIP is the threat that disaster relief will be withheld in nonparticipating or suspended communities. There has been no major study relative to this provision, but FEMA believes that it is being enforced, because of its economic value, even by federal assistance agencies that have sometimes shown hostility to environmental types of restraint, for example, the Small Business Administration exempts loans under $300,000 from Executive Order #11988 which limits federal investment in flood-prone areas (FEMA, 1983). The real problem is that disaster relief continues to be available for such things as public infrastructure and lost crops: any ban is limited to federal assistance related to NFIP-insurable structures. This issue is not a priority in coastal barrier communities, however, since most have joined the program.

The federal government now pays only 75% of disaster-related costs for public assistance to community and state properties (schools, public buildings, roads, dikes, sewers, etc.). The state and local governments must provide the other 25% or lose the federal benefits. This could potentially promote enforcement of NFIP building regulations and serve related purposes. For example, Florida has considered barring use of its 25% funding for construction of facilities that promote growth (Florida DCA, 1983).

Despite the general prohibition against disaster relief payments to landowners in participating communities who do not buy flood insurance, an exception has been made since 1974. After a disaster, these individuals can obtain disaster relief if they buy a flood insurance policy and remain in the program. The cost of the first premium may be loaned to the truly needy as part of disaster relief. This provision lowers the inducement to buy insurance ahead of time but should not cause a problem if it is limited to a one-time exception. A monitoring system is necessary to ensure that the insurance is maintained and the exception is not abused.

As already stated, studies have concluded that the availability of flood insurance may induce development, at least marginally, in the rather uncommon communities in which banks previously refused to insure buildings in some flood-prone areas (e.g., Galveston, Texas, and Rhode Island). In view of the inability of FEMA to enforce tougher regulation of building standards, the General Accounting Office and others have repeatedly suggested that some high-hazard areas be declared build-at-your-own-risk areas, where no subsidized insurance would be available. This was accomplished in designated undeveloped coastal barriers by passage of the Coastal Barrier Resources Act of 1982. It has been suggested that, after major storms, certain developed coastal barrier areas could be placed under the coverage of that act and thus declared ineligible for further insurance.

Sheaffer and Roland (1981) estimated that the federal subsidy to coastal barrier development averages $53,000 per developed acre. The American Red Cross (1977) figures for flood losses are also high. It has been proposed that the federal government buy storm-damaged properties as an alternative measure. Nevertheless, FEMA administrators believe that it is not politically defensible to spend limited funds to acquire these properties, many of which are second homes, even if it would save the federal government money in the long run.

Concluding Recommendations

1. FEMA regulations should ban any alteration of dunes, either within or outside of V-zones (other than restoration and revegetation), since dunes in their natural state are essential to preventing erosion.
2. Erosion-damage coverage should be dropped, unless communities adopt development setback lines for erosion anticipated within a minimum of 30 years. The present coverage encourages claims for natural and inevitable loss of poorly sited structures.
3. Regulations should establish minimum building setback lines at least to the highest high tide line (instead of mean high tide). If coverage is continued for buildings which "migrate" forward of the setback line, use of proceeds to rebuild in that area must be banned.
4. Solid walls under elevated structures should be banned. The temptation to substitute fixed walls for the required "breakaway" walls is too high.
5. The definition of "substantial improvement" should be altered to require new construction standards if damage *or* repair cost—whichever is higher—exceeds 50% of pre-damage market value. Enforcement should be stepped up. This change would limit abuse of the provision that avoids application of the elevation standards for new construction.
6. Casualty losses for uninsured structures in coastal hazard zones (Zones A or V) should be dropped from the Internal Revenue Code. Such deductions encourage and subsidize building in dangerous areas.
7. Private insurance inducements, such as surcharges on policies in communities placed on probation for non-enforcement, should be used more widely, to encourage pressure from within the community to stay in the program.
8. Future federal and state investments in V-zone areas with bad storm histories should be cut back as far as politically feasible.
9. In delineating flood hazard zones, FEMA should take into account the probable future development of an area, since new construction often increases storm hazards.

Acknowledgments

The author wishes to express her appreciation to the following individuals who provided information of value to this paper: Edward Thomas, FEMA; Michael Robinson, FEMA; Michael Bischara and Sheldon Shapiro, Mass. DEQE; H. Crane Miller, Esq.; Jon A. Kusler, Esq. Conclusions and recommendations expressed in this paper are solely the responsibility of the author and do not necessarily reflect the views of FEMA staff or others whom the author interviewed.

References

American Red Cross. *Information and Statistics: Data on Dwellings Destroyed and Damaged . . .1969-1976*. Washington: unpublished mimeo, 1977.
Association of State Floodplain Managers. 1982. *Model State Legislation for Floodplain Management*. Madison: ASFPM.
Association of State Floodplain Managers. 1983. *Preventing Coastal Flood Disasters: The Role of the States and Federal Response*. Special Publication no. 7. Boulder: Natural Hazards Information Center, University of Colorado.
Federal Emergency Management Agency. 1980. *Alternatives for Implementing Substantial Improvement Definitions*. Washington: FEMA.
Federal Emergency Management Agency. 1983. *National Flood Insurance Program and Multi-Hazard Conference*. Summary Report. (Mimeo).
Federal Emergency Management Agency. 1984. "Watermark" (Summer).
Federal Insurance Administration. 1977. *Proceedings of the National Conference on Coastal Erosion*. Washington: FIA.
Florida Department of Community Affairs. Unpublished memo to Interagency Management Committee, dated July 18, 1983.
Gibson, J.M. 1984. "The Future of FEMA's Floodplain Mapping Program." In *Managing High Risk Flood Areas: 1985 and Beyond*. Ed. Monday and Butler. Special Publlication no. 11. Boulder: Natural Hazards Information Center, University of Colorado, pp. 155-166.
IEP, Inc. *Reducing Flood Damage Potential at Ocean City, Maryland* (Preliminary draft, 1983). Wayland, Mass.: unpublished mimeo.
Kunreuther, H. 1978. *Disaster Insurance Protection: Public Policy Lessons*. New York: Wiley Interscience.
League of Women Voters. 1982. "The National Flood Insurance Program: Is It Working?" Washington: LWV.
Massachusetts Division of Water Resources. 1983. *Developed Barrier Beaches: A Coastal Survey*. Boston: Mass. Coastal Zone Management Office.
Miller, H.C. "Enforcement of Substantial Improvements Regulations: Scituate, Mass." Washington: Federal Insurance Agency, unpublished mimeo, 1979.
Miller, H.C. 1983. *Hurricane Alicia: Learning from Galveston*, Report to Office of Ocean and Coastal Resource Management. Washington: mimeo.
Miller, H.C. Undated (a). "Coastal Floodplain Management and the National Flood Insurance Program." HUD contract no. H-4094, unpublished mimeo.
Miller, H.C. Undated (b). "Coastal Flood Hazards and the National Flood Insurance Program." HUD contract no. 1442-17, unpublished mimeo.
National Research Council. 1984. *Hurricane Alicia: Galveston and Houston, Texas, August 17-18, 1984*. Washington: National Academy Press.
Platt, R.H. 1978. "Coastal Hazards and National Policy: A Jury-Rig Approach." *Journal of the American Institute of Planners* (April), pp. 170-180.
Platt, R.H. 1985. "Congress and the Coast." *Environment* (July-August), pp. 12-17, 34-40.
Platt, R.H., and G.M. McMullen. 1980. *Post-Flood Recovery and Hazard Mitiga-*

tion: Lessons from the Massachusetts Coast February 1978. Amherst: Water Resources Research Center, University of Massachusetts. (Available from NTIS, #PB-207 517.)

Ralph M. Field Associates, Inc. 1983. *Preparing for Hurricanes and Coastal Flooding: Handbook for Local Officials*. Washington: FEMA.

Sheaffer and Roland, Inc. "Barrier Island Development Near Four National Seashores." Unpublished report prepared for CEQ and FEMA, 1981.

U.S. Department of the Interior. 1985. *Final Environmental Impact Statement: Undeveloped Coastal Barriers and Federal Flood Insurance*. Washington: DOI.

U.S. General Accounting Office. 1982. *National Flood Insurance: Marginal Impacts on Floodplain Development: Administrative Improvements Needed*. GAO/CED-82-105. Washington: GAO.

Appendix

NFIP Regulations for Coastal High Hazard Zones (V-Zones)
44 *Code of Federal Regulations* sec. 60.3 (e) as amended
in *Federal Register* of September 4, 1985, pp. 36024-5

(e)When the Administrator has provided a notice of final base flood elevations within Zones A1-30 and/or AE on the community's FIRM and, if appropriate, has designated AH zones, AO zones, A99 zones, and A zones on the community's FIRM, and has identified on the community's FIRM coastal high hazard areas by designating Zones V1-30, VE and/or V, the community shall:

(1) Meet the requirements of paragraphs (c)(1) through (c)(11) of this section;

(2) Within Zones V1-30, VE, and V on a community's FIRM, (i) obtain the elevation (in reltion to mean sea level) of the bottom of the lowest structural member of the lowest floor (excluding pilings and columns) of all new and substantially improved structures, and whether or not such structures contain a basement, and (ii)maintain a record of all such information with the official designated by the community under 59.22(a)(9)(iii);

(3) Provide that all new construction within Zones V1-30, VE, and V on the community's FIRM is located landward of the reach of mean high tide;

(4) Provide that all new construction and substantial improvements in Zones V1-30 and VE, and also Zone V if base flood elevation data is available, on the community's FIRM, are elevated on pilings and columns so that (i) the bottom of the lowest horizontal structural member of the lowest floor (excluding the pilings or columns) is elevated to or above the base flood level; and (ii) the pile or column foundation and structure attached thereto is anchored to resist flotation, collapse and lateral movement due to the effects of wind and water loads acting simultaneously on all building components. Wind and water loading values shall each have a one percent chance of being equalled or exceeded in any given yea (100-year mean recurrence interval). A registered professional engineer or architect shall develop or review the structural design, specifications and plans for the construction, and shall certify that the design and methods of construction to be used are in accordance with accepted standards of practice for meeting the provisions of (i) and (ii) of this paragraph.

(5) Provide that all new construction and substantial improvements within Zones V1-30, VE, and V on the community's FIRM have the space below the lowest floor either free of obstruction or constructed with non-supporting breakaway walls, open

wood lattice-work, or insect screening intended to collapse under wind and water loads without causing collapse, displacement, or other structural damage to the elevated portion of the building or supporting foundation system. For the purposes of this section, a breakaway wall shall have a design safe loading resistance of not less than 10 and no more than 20 pounds per square foot. Use of breakaway walls which exceed a design safe loading resistance of 20 pounds per square foot (either by design or when so required by local or State codes) may be permitted only if a registered architect certifies that the designs proposed meet the following conditions:

(i) breakaway wall collapse shall result from a water load less than that which would occur during the base flood; and,

(ii) the elevated portion of the building and supporting foundation system shall not be subject to collapse, displacement, or other structural damage due to the effects of wind and water loads acting simultaneously on all building components (structural and non-structural). Maximum wind and water loading values to be used in this determination shall each have one percent chance of being equalled or exceeded in any given year (100-year mean recurrence interval). Such enclosed space shall be useable solely for parking of vehicles, building access, or storage.

(6) Prohibit the use of fill for structural support for buildings within Zones V1-30, VE, and V on the community's FIRM;

(7) Prohibit the placement of mobile homes, except in existing mobile homes parks and mobile homes subdivisions, within Zones V1-30, VE, and V on the community's FIRM; (8) Prohibit man-made alteration of sand dunes and mangrove stands within Zones V1-30, VE, and V on the community's FIRM which would increase potential flood damage.

FINANCING COASTAL BARRIER INFRASTRUCTURE

H. Crane Miller

Attorney at Law

Washington, D.C.

Introduction

Evaluation of federal policy, law, and programs for undeveloped coastal barriers is enhanced by understanding how already developed barriers finance their infrastructure needs. How dependent is coastal barrier development on federal assistance to finance infrastructure (e.g., roads, bridges and causeways, water supply and distribution systems, wastewater management, and so on)? Historically, the *initial* stages of coastal barrier development were financed through private sources and tax-exempt state or local debt instruments. The Coastal Barrier Resources Act of 1982 withdrew most direct financial assistance for development on units in the Coastal Barrier Resources System. If federal assistance is available, what impact might that have on future development of the Coastal Barrier Resources System? If federal assistance is unavailable, what impact might that have on future development of the Coastal Barrier Resources System? Some insights are available through the means used to finance community infrastructure on the developed coastal barriers.

This paper reports the author's findings from a survey of several developing coastal barrier communities from Maryland to Texas and their use of traditional and innovative nonfederal methods for financing infrastructure. The survey indicates that where pressure for development or redevelopment is strong, private and nonfederal public means to finance infrastructure will be found.

Financing Infrastructure in Galveston, Texas

Before Hurricane Alicia struck Galveston in mid-August of 1983, new construction on the island was proceeding at an annual rate of $30 million. The City was committed to extensive, second-home residential development on the western 20 miles of the island, in large measure for needed property tax revenues. Several new condominiums and motels immediately behind the seawall were recently completed or were nearing completion when the storm struck.

With winds of 90-115 mph, Hurricane Alicia was a low Category 3 storm, mild in comparison to the devastating 1900 storm that killed at least 6,000 people on the island. Nevertheless, high wind damages were experienced, particularly on the west end of the island near the track of the storm. Local building practices and the absence of an adequate building inspection program were particularly implicated in the damage and destruction that ensued.

In the immediate aftermath of the hurricane, Galveston was faced with a potential financial crisis. The concern was that some owners, burdened with high repair, cleaning, and rehabilitation costs resulting from the storm, might delay or avoid paying taxes, possibly reducing tax revenues due in October to 93% of projections.

A second concern was the impact on the 1984 tax year. The Tax Department reported to the City Council that assessed values might be reduced from 10-16% because of storm damages, possibly reducing property values from $160 million to $256 million.

Finally, the City Council had imposed a $0.70 per $100 valuation tax-rate ceiling; the 1983 tax rate was set at $0.66. If property assessments were reduced as much as the Tax Department estimated, the implied revenue loss could not be made up unless the tax rate were increased above the current $0.70 rate ceiling.

Experience showed otherwise. Tax collections in October 1983 were at an all-time high—greater than 99%—amounting to $300,000 more than the city had projected. For both 1983 and 1984, tax delinquency was reduced, providing greater collections than anticipated. Andrew Carr, Chief of the Tax Department, stated that he believed that the influx of cash from wind and flood insurance payments was instrumental during both years in the enhanced collections.

Secondly, the loss in assessed value proved to be less than half of the initial low estimate. Instead of a 10-16% reduction in assessed valuation, the city experienced a 4.4% reduction. The 1984 assessment increased the island property valuation by 10%; a 15% increase was expected for 1985.

The City Council also increased the 1984 tax rate to $0.685 per $100 valuation (from $0.66), and was expected to keep that rate for 1985. The combined effect of a smaller reduction in assessed value than expected, reduced delinquency, increased valuation for 1984, and an increased tax rate averted the expected fiscal crisis. One year after the storm, one could barely find fiscal or physical evidence of damage from the storm.

The storm was catalytic in other changes that took place. Given the widespread damage and destruction, the city worked closely with developers to encourage large developments in Galveston. One measure of the effort is that new construction increased from $30 million per year to $150 million and was expected to continue at the latter rate through 1986. The new construction was primarily in condominiums and motels. For 1985, Galveston expected to have a

300% increase in the number of motel rooms. And in 1985, the city expected a total negation of the adverse effects of Hurricane Alicia.

Major factors in the new growth include the City Council working closely with developers and encouraging large-scale development and use of innovative financing under the Texas Tax Increment Financing Act of 1981 (Article 1066e, Vernon's Texas Civil Statutes). By September 1984, one year after Hurricane Alicia, six new "Reinvestment Zones" had been created by the City Council, bringing to nine the total of such zones in Galveston.

Developers interested in tax increment financing work with the city to designate a proposed development area as a Reinvestment Zone. The developer prepares a project plan for the Reinvestment Zone, describing the public improvements to be made (e.g., roads, stormwater drainage, lighting, sewers, wastewater treatment, water distribution lines, and so on), and makes a firm estimate of the total costs of the project. The plan then is presented to the City Council for approval.

The city enters an agreement with the developer whereby the developer undertakes to construct and finance the project, and the city undertakes to accept title to the public facilities upon completion, the developer retaining an equitable interest pending payment by the city. The parties fix a tax increment base for the property to be developed, that is, fix a tax assessment value for the property in its predeveloped state. All increases in the assessed value of the project property over and above the tax increment base are for the benefit of the reinvestment zone.

All taxes levied against the increased value are deposited into a tax increment fund. Deposited monies are paid monthly to the developer against the total project cost, withholding certain portions for administrative expenses, police protection, and a form of impact payment to the Galveston Independent School District. The city's agreement to pay for the project is expressly limited to the tax increment fund established for the particular reinvestment zone; the full faith and credit of the city is not pledged.

The terms of the agreements range with the size of each project; 10- to 14-year terms are not uncommon. During that period the city also agrees to pay interest on all reimbursable expenditures by the developer.

Galveston's use of tax increment financing is unique in the author's experience. First, Texas's Tax Increment Financing Act contemplates, in common with similar laws in other states, that its authority will be applied to redevelopment of blighted areas of a community. Galveston is using the authority mostly to foster development on previously undeveloped land. This use of the authority apparently has not been challenged in the courts. Secondly, while coastal communities in Florida, North Carolina, and other parts of the coast are increasingly shifting the cost of new public infrastructure to the consumer through the developer, Galveston is unusual in its willingness to reimburse the developer for the infrastructure.

Galveston is using state tax increment financing more than any other community in Texas. Like many other communities, Galveston is near its debt limits for general obligation bonds, and it appears that its voters are unwilling to increase their property tax rates. In August 1984, a referendum on a proposed $40 million bond issue for street repairs and water and sewer improvements failed to pass.

Industrial Development Bond (IDB) financing has been used extensively in downtown Galveston to foster renovations of apartments and commercial businesses on "the Strand." No local referenda or state ceilings apply to IDB financing, only federal Internal Revenue Code limits. At this time, Galveston exceeded its pro rata limits for IDB financing under the federal allocation process. Hence, this form of financing was not available for the city to encourage the growth that was going on.

There are a number of advantages and potential problems with Galveston's use of TIF. Indebtedness under the Texas Tax Increment Financing Act is not subject to the debt service limitations of the state, which apply to general obligation debt. The advantage is that TIF is available at a time when the city wishes to encourage large-scale development and could not otherwise finance publicly the necessary infrastructure. The disadvantage is that the legislature could move to include the TIF debt in the tax base, subjecting it to the state's 8% annual tax base increase limitations and tax referenda requirements, as well as to the possibility of a voter initiative for a rollback.

Gulf Shores, Alabama

In September 1979, Hurricane Frederic made landfall on Dauphin Island, Alabama, causing very serious damage. Gulf Shores was one of the communities devastated by the storm. In the wake of the storm, local officials saw the widespread destruction as an opportunity to build anew in a way that perhaps was politically impossible before the storm. Concerned about shoddy commercial development in the central business area, the storm gave officials an opportunity to attract higher quality, more expensive development than before.

Just as Galveston underwent a building boom following a major storm, Gulf Shores began rebuilding in the immediate aftermath of Frederic. That growth has continued for six years. In the first year after the storm, over 300 single-family homes were rebuilt. Motels and condominiums in the downtown area were repaired and occupied in that first year, as the community experienced a building boom while the rest of the country was in a recession.

The growth continues today as both the East Beach and West Beach areas are being developed with motels and condominiums. Unlike Galveston, however, Gulf Shores has not turned to innovative financing to meet its infrastructure needs. Rather, it has used traditional general obligation bonds, a form of impact fee, and, recently, privatization to handle the growth.

General Obligation Bonds

Following the storm, Gulf Shores borrowed to upgrade its water and sewer lines and repay a federal disaster loan. It did this with a $4.2 million general obligation bond issue, authorized by the Mayor and City Council. Alabama is unusual in the degree to which local government is authorized to pledge its full faith and credit without a voter referendum, and none was sought in this instance. Similarly, the Mayor and Council approved general obligation warrants in 1983 and 1984 without referral to local voters.

Sewer and Water Fees

Gulf Shores has insufficient sewage capacity to handle the growth that it is experiencing. To pay back the general obligation warrants with which it is paying for the expanded capacity, Gulf Shores has a schedule of sewer and water fees, with an added surcharge for new sewer hookups.

Privatization

As of November 1984, Gulf Shores was moving toward privatization of its sewer system. It was in the process of negotiating with an engineering firm to take over the ownership and operation of the sewer system, a move that is found with increased frequency throughout the United States. Funds earmarked for sewer projects by the town would go toward payback of the company.

Financing Infrastructure in Florida

Despite all the development taking place in Florida, that state ranks among the 10 lowest states in level of general obligation bonded debt. This reflects the history of bankruptcy of many Florida municipalities during the Great Depression, as well as an elderly voting population that regularly defeats general obligation bond referenda for fear of increased real property taxes. There are notable exceptions, such as in Sarasota County and Dade County, but most Florida counties do not commonly use general obligation bonds to finance their public infrastructure. Instead, a variety of other means are used, including privatization, revenue bonds (including tax increment financing), and impact fees.

Sanibel, Florida

The City of Sanibel relies to a great extent upon privatization for much of its infrastructure needs. Its sewage disposal, water, and fire facilities are privately owned, financed, and operated. The Island Water Association, for instance, is a cooperative which financed its major capital needs through $5 million in borrowings from the Farmers Home Administration from 1965 to 1979. It has obtained no loans since 1979 but has financed improvements through water charges, which include an allowance for depreciation, and through connection fees. The $1,825 connection fee for single-family homes and $1,531 fee plus the cost of a meter for each dwelling unit of condominiums, for example, are comparable in cost to those charged in Gulf Shores, Alabama, and are apparently adequate to cover the costs of service, meter, and plant expansion included in the fee.

Sanibel's sewage disposal system was initially financed by the original developer of the island, from whom the current operators assumed responsibility for the short-term construction loans used to finance the system. Subsequent financing for expansion of the system came from loans from local lenders, secured by the income stream from the operation, which is also subject to regulation by the Florida Public Service Commission.

Most of Sanibel's local streets were built by developers. The main road to and on the island is county owned. Monies for maintenance, repair, and improvements of the island's road system come from many sources: causeway toll income, a vehicle weight permit fee imposed by the city, state payments in lieu of taxes on the J.N. "Ding" Darling Wildlife Refuge, and revenues received from an interlocal agreement with Lee County.

Impact Fees

Impact fees are seen by many as the answer to Florida's financing of infrastructure. However, the experience to date suggests that neither state nor local government can charge enough through impact fees to meet their entire capital infrastructure needs, particularly regarding transportation. For instance, a fiscal model for Sarasota County and other local governments estimated that impact fees of $15,000 per dwelling unit were necessary to cover the impacts *outside* any given development project. A major issue for Florida is not whether developers finance and provide infrastructure within a development, but the extent to which they pass on to consumers the cost of impacts outside the development. Impact fees are assessed on a relatively small base—new construction—and are not keeping abreast of the needs.

Litigation challenging the imposition of impact fees has delayed the acceptance of those fees. However, as courts have ruled on challenges of fees for water and sewage facilities, parks, and roads, greater use of this approach is now found throughout the state.

An important coastal barrier example is found in St. Lucie County, on Hutchinson Island. Faced with development pressure, environmental sensitivity of extensive mangrove, wetland, and dune areas on North and South Hutchinson Island and the inadequacy of their roads and bridges to evacuate the population in case of a hurricane or other emergency, St. Lucie County created the Hutchinson Island Residential District (HIRD).

HIRD is a zoning measure that seeks to ensure that residential development beyond the existing or base level will not take place until certain levels of vehicular traffic can be accommodated. The ordinance classifies lands into three geographic districts. Four maximum residential density levels are established for each district, expressed as a percentage of the maximum density set forth in the St. Lucie County Growth Management Policy Plan. To increase development densities beyond the percentages in the density table, measures to increase traffic capacity are specified in the ordinance. They range from traffic signals and an additional right-turn lane on the north island at a cost of about $200,000, to construction of a two-lane bridge to South Hutchinson over two miles of water, estimated to cost $45 million.

Provision is made in the ordinance for construction and use of structures at a density greater than the maximum set in the density table upon payment of an alternate development fee, payable in advance on issuance of building permits and applicable to each unit exceeding the maximum.

This regulatory concept, worked out as a compromise among development, environmental, and local government interests, has yet to be tested in the soft real estate market. Currently there is excess residential capacity on Hutchinson Island, and no demand to build beyond the base level. Moreover, Hutchinson Island may be designated by the state as an "area of critical state concern." Such designation could restrict development on the island to one dwelling unit per acre throughout the entire island.

Tax Increment Financing

While authorized by state legislation in the 1970s, no successful application of TIF existed in Florida as recently as March 1982. Since then, TIF has been used in Lakeland, Orlando, Pensacola, and St. Petersburg to finance redevelopment of "blighted" areas within those cities. This innovative approach is seen as an increasingly important means to obtain public financing for redevelopment projects, particularly in communities where voters have historically rejected referenda on general obligation bonds, or where there are few other options for financing downtown redevelopment.

The Florida procedure is relatively simple to establish and administer. Cities are required by statute to give notice to other taxing authorities of the proposed use of tax increment financing, but do not have to obtain their approval. The initial assessed value of a designated community redevelopment project area is determined at the time a city adopts a comprehensive community redevelopment plan.

The portion of the real property taxes attributable to the initial, assessed value is allocated to various taxing districts (counties, cities, road, park, health, fire, and other municipal corporations or districts with power to levy property taxes in the redevelopment project area). School districts, water management districts, and library districts have been specifically exempted by statute from limitations of the initial assessed value.

The portion of the real property taxes attributable to the increase in current, assessed valuation of real property in the project area over and above the initial, assessed valuation, is paid into a special redevelopment trust fund. Monies in the fund can be used to pay principal and interest on bonds, loans, advances of money, or other indebtedness to finance the redevelopment project.

There is evidence that some developers are pressing for authority to use tax increment financing for new development, as in Galveston. This trend may well be opposed by other taxing districts whose revenues from TIF project areas would be curtailed. As this form of financing increases, we can expect that more special districts with property taxing powers will seek to be excepted from the limitations, as have Florida school and library districts. The battle lines are drawn.

Financing in North Carolina

As in Florida, many municipalities in North Carolina became bankrupt during the Great Depression, and out of that experience has evolved state law governing the budgeting and fiscal control of local governments. The Local Government Budget and Fiscal Control Act, N.C.G.S. Sec. 159, deals not only with the annual budgeting and expenditure limitations on local government, but also controls local governments' access to financial markets for capital expenditures.

The act requires local governments to adopt and administer an annual balanced budget. It imposes a net debt limitation of 8% of appraised property values in the locality, except in the case of certain capital projects.

By the act, local governments are expressly forbidden to go to private sources for debt obligations; they may incur no indebtedness except through the North Carolina Local Government Commission (LGC). When a local government needs to sell bonds for public infrastructure, the LGC works with them to assess the project, assess the locality's ability to service the debt, discuss financing options (e.g., general obligation vs. revenue bonds), set up a debt service schedule, obtain a bond rating, and market the bonds.

The approach of the state, and especially the LGC, has been conservative. As a result, one finds little financing innovation in North Carolina's coastal communities and a general perception that the ad valorem property tax works adequately. According to state officials consulted, none of the coastal communities is currently approaching its statutory property tax cap. General obligation bonds are most commonly used for capital expenditures; revenue bonds and tax anticipation notes are not heavily used in coastal areas of the state.

Nags Head

The Town of Nags Head exemplifies the general characteristics of coastal community financing noted above. The town issued general obligation bonds for its first water system, has undertaken no general indebtedness since, according to the town manager, and is not near its statutory indebtedness limits.

The general procedure for subdivision development is for the developer to build roads, water lines, and so on, according to town standards. Upon completion, the town accepts dedication of the roads for maintenance purposes.

Outside the developments, a graduated water impact fee, based on the size of the development, is imposed on new construction. Town assessments for road improvements are authorized by the state. Throughout the coastal communities, impact fees are being used widely, imposing fees at the front end of new construction to cover the costs of water lines, expansion of water systems, and so on, which the developer adds to the capital cost of the structure at the time of sale.

Nags Head uses no innovative financing. It has used no revenue bonds, lease purchase, sale leaseback, or techniques other than the traditional methods outlined to meet its needs. It has received federal loans and grants for its water system in the past, is a recipient of state aid for roads and other facilities, and shares with Dare County part of the proceeds of a $0.01 local option sales tax, among the sources of revenue other than the property tax.

By all accounts, Nags Head is in healthy financial condition. The recent sale of the 400-acre Epstein tract in the middle of Nags Head will bring considerable new development to the town. It appears that the town will be able to meet its expansion needs using the traditional financing methods that have served it in the past.

Ocean City, Maryland

With $175 million in building permits in 1983 and $80 million in 1984, Ocean City, Maryland is undergoing vigorous rebuilding of its downtown area, while continuing to expand toward its northern boundary. According to the City Manager, the city has had no problems either raising money or obtaining voter approval of general obligation bonds. For instance, a $6 million obligation bond proposal was approved in November 1984. No leaseback, tax increment financing, or other innovative financing appears needed, understood, or approved by the City Council, which has met the city's public financing needs wholly by traditional means.

As measure of the development and vigor of the local economy, the city reduced its tax rate and held the rate at the new level for a second year, under a policy not to increase the tax rate. It has increased license and building permit fees, does not use dedication or exaction techniques (although it does require a public beach easement for shoreside development), and it does not impose impact fees.

In common with many other coastal communities with small permanent populations and large summer populations, Ocean City has virtually no poverty or racial minorities and does not qualify for major revenue sharing or Department of Housing and Urban Development assistance.

Ocean City is a picture of vigorous development and fiscal conservatism. Its local economy is tied to the vagaries of tourism, the national economy, the price and availability of fuel, and similar factors which it shares with other coastal communities economically oriented to a single season of the year. It is clear also that its economic development is not dependent upon federal or state aid, and as long as the City Council continues to encourage development and the national economy does not decline, the city will continue to attract new development and will have little problem financing the urban infrastructure needed to support that development.

Conclusions

The growth being experienced in several of the communities reviewed here is attributable to factors such as national economic recovery, lowered interest rates, innovative financing occasionally, traditional financing in almost all instances, and the strength of the local real estate market. In the one instance cited (Hutchinson Island, Florida) of a soft real estate market with surplus units available, further development has virtually stopped. Tax and expenditure limitation systems and reduced federal aid appear to have relatively little to do with the volume of development going on but have much to do with the adaptations to those changed conditions that have been made.

Interviews and Sources

Galveston, Texas
    Andrew Carr, Chief, Tax Department, Oct. 1, 1984.
    John Miles, General Counsel, Texas Property Tax Board, Austin, Texas, Oct. 11, 1984.
    Miller, H.C. *Hurricane Alicia: Learning from Galveston.* Report to the National Oceanic and Atmospheric Administration, U.S. Department of Commerce, October 1983.

Gulf Shores, Alabama
    Nancy Hollingsworth, Treasurer, Nov. 1, 1984.
    Donald Howell, Town Clerk and Administrator, Nov. 1, 1984.

Florida (general)
    James Nicholas, Acting Director, Institute at Florida Atlantic, Nov. 1, 1984.
    James Quinn, Chief, Bureau of State Planning, Oct. 22, 1984.
    Neil Sipe, Bureau of Economic and Business Research, University of Florida, Gainesville, Dec. 28, 1984.

Sanibel, Florida
    Mildred Howze, Finance Department, Oct. 30, 1984.
    Robert Duane, Planning Department, Oct. 30, 1984.
    Ray Pavelka, Vice President, Mariner Properties, Inc., Ft. Myers, Dec. 27, 1984.
    Robert E. Hollander, General Manager, Island Water Association, Dec. 27, 1984.

Lakeland, Florida
    William Hill, Executive Director, Lakeland Downtown Development Authority, Dec. 19, 1984.

Orlando, Florida
    Thomas Kohler, Executive Director, Orlando Downtown Development Board, Nov. 1, 1984.

Pensacola, Florida
    Andrew Hamm, Executive Director, Community Redevelopment Agency, Nov. 1, 1984.

St. Petersburg, Florida
    William Wardwell, Office of the Assistant Administrator of Community and Economic Development, Nov. 2, 1984.

St. Lucie County, Florida
    Dennis Murphy, Assistant Planner, Board of County Commissioners, Dec. 28, 1984.

North Carolina (general)
    David Owens, Director, Office of Coastal Management, North Caro-

lina Department of Natural Resources and Community Development, Oct. 17, 1984.

Winn Quakenbush, Local Government Commission, North Carolina Department of the State Treasurer, Oct. 17, 1984.

Spencer Rogers, North Carolina Sea Grant Program, Oct. 17, 1984.

Nags Head, North Carolina

Webb Fuller, Town Manager, Oct. 17, 1984.

Doris Guard, Finance Department, Oct. 17, 1984.

Ocean City, Maryland

Anthony Barrett, City Manager, Oct. 17, 1984.

MANAGEMENT ALTERNATIVES

A TIME FOR RETREAT[1]

Orrin H. Pilkey

Duke University

Introduction

I would like to present a scientist's view of the coastal crisis, particularly the crisis on our recreational beaches. A good example is the saga of Sea Bright, New Jersey. In March of 1984, this coastal village was attacked by a significant northeaster, probably a ten-year storm. Immediately after the storm, Sea Bright claimed, at least for the benefit of *The New York Times,* total damages of $82 million. But none of the buildings in town were damaged to any significant degree. Furthermore, the total value of all the buildings in town two years before was estimated to be only about $65 million.

There is no question that the damage estimate figure was somewhat exaggerated. Nevertheless, Sea Bright can be considered to be the American endpoint in shoreline stabilization. Here is a community that received very significant financial damage in a minor storm. The damage was even significant in terms of the total value of all the buildings, and yet none of them were destroyed. All the damage occurred on the seawall and the beaches, according to the newspaper articles. How many more storms can such a community withstand?

Recent Advances in Coastal Research

During the last two decades, scientists have made a great deal of progress in understanding shorelines and geological processes of the shoreline. Perhaps the most significant single discovery is our now very well-documented understanding of the migration of barrier islands (Swift, 1975). Twenty years ago, migration of

---

[1] This paper is based on a dinner speech presented at the "Cities on the Beach" conference on January 16, 1985.

barrier islands was not heard of or even thought of, except by a few far-seeing individuals whom nobody believed. Another major accomplishment or fallout of recent research in the last couple of decades is the recognition of sea level rise. We have documented a very important fact, namely, that sea level rise has accelerated in recent decades (Revelle, 1983). It is now rising on the order of one foot per century, and it is probably caused by the greenhouse effect. The Council on Environmental Quality informed the Executive Branch that our society faces a number of problems because of the greenhouse effect. The National Academy of Sciences informed the Legislative Branch of the same problems. And the Environmental Protection Agency informed all of us that the sea level rise would be somewhere between two and ten feet by the year 2100 (Barth and Titus, 1984).

We still have much to learn. Perhaps the most gaping hole in our understanding of the coast has been the long-range impact of various stabilization schemes on shorelines. There are many things that need to be documented, that need to be quantified, and many important things that we do not really understand about stabilization. Also, day-to-day and storm dynamics of coastal barrier shorelines remain a mystery, especially the "oceanography" of the shoreface. For example, we have not begun to quantify sand supply or sand sources as much as we should be able to. Also, we do not fully understand inlets. Much progress has been made on tidal delta morphology and sand transfer mechanisms. But we are a long way from understanding inlets to the extent that we can predict or mitigate some of the negative impacts that they have on human developments.

Recent Changes in Attitudes and Management

Nonetheless, as they say, we have come a long way. Simultaneous with our real advances in the understanding of shorelines, many other things have been happening. Most important, there has been an incredible amount of development along (and often right next to) our sandy shorelines, our coastal barrier shorelines. It is now recognized that erosion is affecting all shorelines in the world. Perhaps, to one degree or another, that is not quite true. A better figure would be somewhere between 80 and 90 percent, depending on whom you believe and how you define erosion. Since 1930, essentially all of our American shorelines have been retreating.

Another change has been an incredible increase in the cost of coastal property; the closer to the surf, the higher the price. For example, a quarter-acre beach lot near Monterey Beach, California recently sold for $500,000.

We have also experienced a change in our attitudes toward erosion. For example, during the 1930s, before we knew about barrier island migration, engineers, geologists, and everyone else agreed that artificial dunes had to be built on the Outer Banks of North California. Otherwise, the islands were going to disappear. We now know that this is not the case.

On developed barriers in the 1950s, if a house was threatened by erosion, what else would one do but build a seawall? It made common sense not to let a house fall in. In the 1980s, the answer is not quite so obvious. Perhaps one good barometer of the prevailing attitude is a discussion that took place at a recent meeting of the Florida Chapter of the American Shore and Beach Preservation Association. It was stated that there are really only three alternative solutions to the erosion problem. There is hard stabilization, which they all agreed was bad. There is beach replenishment. And there is abandonment of the shoreline. Those are the three options. So the obvious choice was beach replenishment. That is an attitude among beachfront property owners that is a fairly prevalent one now, especially in Florida and perhaps along much of the Gulf of Mexico.

Finally, during the 1980s, we are beginning to recognize that beaches are both precious and diminishing. Newspaper editorials more frequently note that seawalls are built to save the property of a very small number of people. The attitude toward beaches is beginning to be something like that toward the national parks. To build a seawall along a recreational beach is now looked upon as something akin to dumping garbage in the Grand Canyon. There seems to be a sudden and profound change in the public attitude toward beaches.

We have only recently recognized the very long range and high cost of any kind of stabilization; that everything we do on the shoreline is temporary. It is also apparent now that replenishment such as can be seen on nearby Virginia Beach is not the answer for the East Coast of the United States. It may be an answer here and an answer there, but there is no way to replenish the entire East Coast. We must find another way.

Prospects for the Future

As the 1980s roll on, we continue to face a fascinating problem, namely what I consider to be the difference between the scientific mentality and the engineering mentality. Geologists are fairly qualitative, as scientists go. And many coastal geologists tend to be more qualitative than other geologists. This is in opposition to coastal engineers who are very quantitative. It is a difficult problem, because the public receives very mixed views regarding the problem of shoreline erosion. If the leaders of a coastal community ask an engineer for a solution to a beach erosion problem, they will get one as fast as they can ask for it. If they ask a geologist for a solution to the problem, they will be told that it is really very complex and there really is no solution to the problem. Of course, the engineer is correct for the 10- to 20-year time horizon and the geologist is correct for the 30- to 50-year time span. It all depends on your viewpoint and your priorities, but the difference between the scientific and engineering mentalities has greatly complicated coastal management.

Taking into consideration the sea level rise, the high cost of stabilization, the fact that erosion is so very widespread, and the fact that environmental damage is the rule when stabilization is used, it boils down to a situation where we can have shorefront buildings, or we can have beaches. But we cannot have both.

Examples of Statewide Initiatives

So, what are the states doing about this? A variety of actions are taking place. One example is in Louisiana, where millions of dollars are being allotted to stabilize as much of the shoreline as possible to keep it from retreating. Most of this is not really destroying recreational beaches. It is mainly an attempt to stop the delta shoreline from retreating, so that they can keep their oil revenues from retreating back to the federal treasury. Louisiana is putting all the resources it possibly can into stopping shoreline erosion, mostly using money that the federal government gives for mitigation of environmental problems caused by offshore oil. There is some irony in the use of such funds for hardening the shoreline and, in any event, no one believes it will work for long.

The other extreme might be North Carolina or Maine, where it now is against the law to build a seawall on a sandy beach. This is a fairly easy thing for Maine to do, since only about one percent of the beaches are sandy. And something like five or ten percent of the people who live near the beach vote in Maine.

North Carolina, however, is in the process of doing something very courageous. They have just established regulations to halt all stabilization on the open ocean shoreline. And I believe they will succeed, but of course such regulations will be tested in the political arena, and the real test will come when the first eight-story condominium is about to fall in.

Not surprisingly, there are a variety of outlooks in California (Griggs and Savoy, 1985). The Baines Bill is still active, but has not yet passed. This bill says that anybody whose house is threatened has a right to build a seawall immediately, to keep it from falling in. On the other hand, the owner of a three-story building at Monterey Beach has been prevented from building a seawall, in spite of the fact that the building will most likely fall into the sea this winter or next. And the reason that this individual was prevented from building a seawall is because it would have damaged the beach.

In Florida, the 1984 Thanksgiving Day Atlantic Coastal Storm caused much reconsideration of the stabilization issue. The northeaster, which damaged essentially no buildings, removed massive amounts of beach sand from Jacksonville down to Fort Lauderdale. Most of its notoriety came from the ship ("Mercedes") that bumped into a seawall in Palm Beach. Whether the 1984 storm will really have any long-range political impact remains to be seen, but it looks encouraging.

What is happening in North Carolina and Maine and, to some degree, in Texas (Morton et al., 1983) basically amounts to a retreat from the shoreline. If you do not allow stabilization of a shoreline; if you do not allow construction of hard structures, the existing buildings will eventually fall in. There is no other possible ultimate outcome, unless you think you can replenish all of our shorelines. So this really amounts to a retreat, albeit one that is unacknowledged. I think it is time to call a spade a spade and to say that we are now retreating from the shoreline. This is the only fair thing to do for those people who own buildings that are now going to be threatened by shoreline erosion and the lack of stabilization. If you agree that the preservation of the recreational beach is a very high priority, I think you would all agree that it comes down to the necessity for a retreat.

Methods of Shoreline Retreat

Now, just how do we go about retreating? This is something of a forbidden topic that really has not yet been discussed. [Since this presentation, a conference was convened at Skidaway Institute of Oceanography to address specifically the retreat question (See Howard, Kaufman, and Pilkey, 1985)]. The kind of meeting that we are having here, which is a unique meeting in itself, is a good start. We also need to support some research on techniques of retreating from the shoreline. Some ideas that come to mind would be the halting of all high-rise construction near the shoreline. Another idea would be to let all the buildings fall in as their time comes. This would be politically difficult and, for most shoreline managers, it probably is not the solution they are going to come up with if they wish to be re-elected. Another good idea was suggested by the Fire Island National Seashore a few years back. The idea is to migrate Fire Island artificially by pumping sand on the back side and by allowing the oceanside buildings to fall in or moving them to the back side. If we let the buildings fall in, it requires that the oceanfront property owners be good sports.

Another possibility would be to use rolling setbacks. Basically, what this entails is that no one is allowed to repair their home, or their building, and when the building's time comes, in terms of its own natural aging process, or when the shoreline catches up—whichever comes first—you demolish the building. In other words, you let people build wherever and whatever they want, but they have just got to get out of the way, somehow, when the shoreline retreats.

There are various ways to pay for retreat from the shoreline. The National Flood Insurance Program could pay for it, but that could create an unacceptable tax drain and possibly a taxpayers' revolt. In Texas, there is a genuine and serious proposal that local communities will now tax themselves to pay for the moving of buildings as the shoreline retreats past them. To help people move their buildings, we should initiate research concerned with how one moves 20-story condominiums away from a beach. In fact, the 20-story condominium seems to represent the real stumbling block to this whole problem. However, a 20-story condominium is no different than a beach cottage when it comes to the problem of maintaining the quality of the recreational beach.

Conclusions

In any event, we need much research on how to retreat from the shoreline. I think it can be viewed not as abandonment of the shoreline, but rather as a means of living with the shoreline. It is a challenging approach to the management of developed coastal barriers that will require some very creative thinking on our part. And people like us are the ones who will have to lead the way in the retreat.

My conclusion is that retreat from the shoreline is both an economic and an environmental necessity. If the next generation's recreational beaches are to be preserved in the short run, immediate retreat is necessary. If beachfront development is to be preserved in the long run, retreat is also necessary. There are no national-scale alternatives.

It would be better to begin an orderly, strategic retreat now than to retreat in tactical disarray later!

References

Barth, M.C., and J.G. Titus, eds. 1984. *Greenhouse Effect and Sea Level Rise*. New York: Van Nostrand Reinhold.

Griggs, G., and Savoy, L., eds. 1985. *Living with the California Coast*. Durham: Duke University Press.

Howard, J.D., W. Kaufman, and O.H. Pilkey. 1985. *National Strategy for Beach Preservation: Proceedings of the Second Skidaway Institute of Oceanography Conference on America's Eroding Shoreline, Savannah, GA 1985*.

Morton, R.A., O.H. Pilkey, Jr., O.H. PIlkey, Sr., and W.J. Neal. 1983. *Living with the Texas Shore*. Durham: Duke University Press.

Revelle, R.R. 1983. "Probable Future Changes in Sea Level Resulting from Increased Atmospheric Carbon Dioxide." *Changing Climate*. Washington: National Academy Press, p. 433-447.

Swift, D.J.P. 1975. "Barrier Island Genesis: Evidence From the Central Atlantic Shelf, Eastern U.S.A." *Sedimentary Geology* 14(1): 1-43.

AN ACQUISITION PROGRAM FOR STORM-DAMAGED PROPERTIES ON COASTAL BARRIERS: THE STATE ROLE

Gary R. Clayton

Mass. Department of Environmental Quality Engineering
Boston, Massachusetts

Introduction

Developed coastal barriers are subject to varying degrees of recurrent storm damage and a corresponding risk for loss of life and property. Despite the alteration of natural features, developed coastal barriers remain dynamic landforms subject to rapid inundation by storm surge or by permanent inundation from erosion and rising sea level. The degree to which developed barriers will change in response to coastal processes is related, in part, to the morphology of the barrier. Relatively wide, high barriers are generally more stable than low, overwash barriers or chenier plains (Leatherman, 1981, p. 8). Floodplain management techniques such as building standards, floodproofing measures, and shore protection structures eventually prove inadequate on barriers which experience wave impacts, storm overwash, and severe erosion.

One measure coastal states may use to avoid the likely prospect of repeated storm damage and costly recovery is to acquire storm-damaged properties. This may be the most efficient, economical, or sometimes the only method for avoiding future storm damage in these coastal high-hazard zones.

An acquisition program is also likely to be one of the most difficult floodplain management techniques to implement because of real or perceived difficulties in obtaining necessary funding and political support. Coastal land prices are unquestionably expensive. Community officials and landowners may also perceive an acquisition program as an unnecessary intrusion into local affairs, as well as representing a negative impact on property tax revenue. In addition to acquisition costs, funds must also be available for demolition and removal of storm debris and damaged structures. Administrative costs (planning, negotiations, appraisals) for implementing an acquisition program, as well as for management of the acquired properties, will also be incurred.

Nevertheless, for coastal states, an acquisition program for storm-damaged properties offers important and distinct advantages over other floodplain management techniques. The state can often play a critical lead role in implementing such a program because of the possible availability of funding and personnel resources. The state role can also be a complement to local acquisition initiatives or federal funding programs. Moreover, acquired land with storm-damaged properties can be used to meet other state or community objectives, including increased open space, recreation, or public access to the coast. In addition to eliminating storm losses on the acquired property, acquisition can improve or restore the natural features of coastal barriers. For example, re-establishing a primary dune system provides a natural, resilient buffer for more landward properties against storm wave energy. Finally, by acquiring storm-damaged properties, past land-use mistakes concerning development on coastal barriers are corrected with a degree of permanency not provided by other floodplain management techniques. The result can be lower premiums for flood insurance holders and lower public costs to cover catastrophic storm losses.

While the permanent evacuation of entire developed coastal barriers is not generally feasible, the prospect of continued storm damage and loss suggests that, at a minimum, the most hazard-prone areas on these barriers be considered for acquisition. The likelihood of continued submergence and erosion of the coast means that developed coastal barriers will inevitably become more hazard-prone in the decades ahead. Prudent public policy would seem to require that an acquisition program for storm-damaged properties be an important element of a comprehensive floodplain management program that also includes land-use planning and regulation, public investments, and public education. This paper outlines some aspects and considerations for establishing a state acquisition program of storm-damaged properties on developed coastal barriers.

General Aspects of an Acquisition Program

An acquisition program for storm-damaged properties must address a variety of planning, legal, management, and funding issues. Priorities must be established to evaluate potential property acquisitions. Relocation assistance programs are often necessary, as well as a plan to remove storm-damaged buildings and other debris. Finally, the administrative structure to implement these and other aspects of an acquisition program must be clearly defined and adequately staffed and funded before the storm occurs.

THE STATE ROLE IN AN ACQUISITION PROGRAM

Site selection criteria must recognize that not all properties on a coastal barrier are equally susceptible to storm damage. Special hazard zones experience high velocity floodwaters, erosion, and/or wave-transported debris. Tidal inlets, for example, are generally the most geologically active sites on coastal barriers (Fitzgerald and Hayes, 1980, p. 2356). Consequently, storm-induced changes of inlet morphology or location on a developed coastal barrier can be potentially catastrophic. Since in these high-hazard locations, other floodplain management methods are unlikely to be effective in the long term, the degree of hazard threat will often be a primary factor in deciding priorities for acquisition. Other important factors to consider in establishing a priority acquisition plan are the location of a property in relation to other parcels already acquired or to be acquired, direct and indirect costs, the extent of community support, the willingness of landowners to sell, and the intended reuse (U.S. Water Resources Council, 1981). In some instances, suitable residences may be appropriate for acquisition and commercial structures suitable for flood-proofing. Developing a data base that identifies sites which historically have been prone to storm damage can also help target selection strategies. Establishing reasonable and consistent site selection criteria that are incorporated into a state floodplain management plan or by a gubernatorial Executive Order alerts and informs the public as to the types of storm-damaged properties intended for acquisition.

A relocation assistance program for implementing temporary building moratoria also requires pre-storm planning. Financial or technical assistance requirements for relocatees from coastal barriers vary from state to state. Public assistance, in whatever form, is one measure which may further community and landowner cooperation and thus ultimately contribute to the success of an acquisition program. Temporary building moratoria may be especially useful after major storms when numerous structures and utilities are severely damaged. A moratorium provides time for the state or local decision-making bodies to determine when, where, and how rebuilding should occur. It also provides the necessary time to proceed with an acquisition program, particularly if many structures have been destroyed or seriously damaged.

Finally, the implementation of any acquisition program requires an effective administrative structure. The state agency assigned this responsibility must have not only the necessary legal authority but the requisite experience with other land management programs. In addition, appropriate staff expertise in law, engineering, and real estate appraisal is also needed to ensure the effective operation of an acquisition program.

Funding Techniques for an Acquisition Program

Coastal lands can be acquired either through voluntary sale or condemnation for a fee simple or less-than-fee simple interest in the land. Generally, the high cost of coastal land requires a variety of acquisition techniques. Setting acquisition priorities and utilizing various cost-saving techniques will maximize limited acquisition funds.

Acquiring fee simple interests in storm-damaged properties at full or below market value is the most common type of acquisition. Less-than-fee interests in lands can sometimes provide a satisfactory way to achieve state floodplain management objectives. This may include, for example, leases, donations, easements, transfer of development rights or mandatory dedication [see Tripp, this volume]. States may also establish land transfer programs by exchanging storm-damaged property for publicly owned property in another location. This can be a cost-effective way to implement an acquisition program if surplus state or federal property is currently available.

Funds for an acquisition program may be obtained from a variety of state or federal government sources. Identifying and securing these funds before the storm strikes will ensure a smoother and more rapid implementation of the program. At the state level, general revenues or bond authorization funds are typically used for an acquisition program. The beneficial impact of these funds can be significantly expanded by integrating state resources with complementary federal programs. For example, the National Flood Insurance Program reimburses policy holders for severe or total property damage losses. Rather than rebuilding on the same hazard-prone coastal barrier, some landowners may willingly relocate if provided a reasonable alternative. The state can offer this alternative by acquiring the property from the willing seller which, even at fair market value, represents a substantial discount from the pre-storm value of the building and land. Moreover, the homeowner under these circumstances is able to recover his real estate investment.

In the past, federal funds have also been used to acquire storm-damaged properties, including Community Development Block Grants, National Flood Insurance Act Section 1362 funds, Water Resources Development Act of 1974 (Section 73) funds, Small Business Administration loan assistance, and HUD Section 407 funds (U.S. Water Resources Council, 1981; Kusler, 1982). In recent years, however, the availability of these funds has been curtailed or even eliminated. Nevertheless, the combination of limited federal funds with other resources may offer an important source of financial assistance for acquisition programs.

Management Issues for Acquired Storm-Damaged Properties

State acquisition of storm-damaged properties on developed coastal barriers leads to land management responsibilities. Initially, this includes the removal of storm debris in order to secure the acquired properties. Several options for debris clearance are available. Acquired buildings that remain structurally sound after a storm can be moved by the landowner or auctioned to the highest bidder for sale and relocation. In many instances, storm activity has resulted in seriously damaged structures which can only be removed by demolition. Demolition requires funds, equipment, and personnel, as well as suitable locations for the disposal of debris. A debris clearance program also involves regrading sections of the coastal barrier to remove foundations, pilings, or cellar holes. The physical modification of the beach and dune area by this regrading represents a critical opportunity to restore some of the former storm buffer capacity afforded by these landforms.

Longer-term management considerations will largely reflect the intended reuse of the property. Passive recreation or open space use of acquired properties may only require minimal maintenance. More intensive recreational activities require greater consideration of the hazard-prone nature of the site. In either case, providing public access to these sites must address not only the availability of parking and protection of restored natural features of the site, but also local concerns associated with new or increased public use of the beach area.

If the state retains title to the land, management of the acquired parcels does not necessarily require the assignment of routine management functions to a state agency. Provisions for community or private management utilizing long-term, low cost arrangements can provide another option. The state can also simply transfer the title to the parcel with appropriate use restrictions to a willing municipality. Either the lease option or the transfer can be effective in managing small, scattered sites on a coastal barrier, especially where the initial implementation of an acquisition program is likely to involve obtaining some non-contiguous parcels.

Management of acquired storm-damaged properties must reflect individual site characteristics and community needs. This may be guided, however, by a comprehensive state management program plan addressing such issues as management agency responsibility, storm debris clearance, site restoration, reuse, and public access. The effective implementation of a management program which is responsive to community interests will help ensure the success of future acquisitions of storm-damaged properties on developed coastal barriers.

The following is a case study of the use of Section 1362 funds to acquire chronically floodprone property at Scituate, Massachusetts.

The Scituate Experience

In 1978, Massachusetts was struck by a devastating "northeaster" winter storm with an estimated return frequency of 75 to 100 years. One coastal community that was especially hard hit was Scituate, a residential community located south of Boston. Scituate is an area that has had a long history of storm damage; accounts of such damage date back to the early 1600s. The 1978 storm was especially severe and resulted in over $7 million in damages to public roads, buildings, and other facilities. Numerous homes were completely destroyed.

Following the storm, immediate efforts began by public agencies to acquire storm-damaged properties for use as public open space. The governor proposed a $10 million state bond issue to buy devastated coastal property for recreational or natural flood control purposes. This program involved only the voluntary sale of storm-damaged property at post-flood fair market value. Public access was required since state funds would be used and relocation assistance for property owners would also be made available. Management plans for acquired properties were also required. Major opposition from coastal residents and officials soon arose, however, following the program proposal. These individuals felt that an acquisition program would adversely affect the property tax base of coastal communities or that the state was simply using the disaster as a cover to obtain public access. The acquisition program was tabled by the state legislature before a vote, as a result of a political compromise which also killed legislation that would have allowed rebuilding without review by municipal commissions.

The use of federal funds for acquiring storm-damaged properties was also considered at this time, but none were available. Not until 1980, when FEMA issued draft guidelines for implementing a pilot program under Section 1362 of the National Flood Insurance Program, did a potential source of funds become available for acquiring storm-damaged properties in Scituate. By the summer of 1980, federal and town officials had identified 20 potential sellers. Although the relevant town boards supported this program, significant citizen opposition remained. When final approval for the acquisition came before a June town meeting, residents of the community, already angered by the state's refusal to construct a proposed seawall, voted it down by a wide margin.

Nevertheless, some property owners as well as town officials remained interested in the 1362 program. The state now approached FEMA to determine if funds were still available. Earlier, the state had chosen not to participate in the acquisition effort, primarily because of the small and scattered nature of the parcels. By August 1980, however, the governor had signed a Barrier Beach Executive Order which directed state agencies to take actions that would avoid or minimize storm-related damage on coastal barriers. In particular, it afforded the highest priority for use of disaster assistance funds for relocating willing sellers from storm-damaged areas.

Shortly thereafter, the state agreed to accept 10 separate parcels consisting of 16 lots and entered into a long-term lease arrangement for the management of these areas with the town of Scituate. Deed restrictions prohibit public facilities or construction of any kind on these lots. Acquisition costs for the 1362 program totalled approximately $500,000.

A number of lessons can be learned from this program's experience. Despite the time that elapsed between the storm and the acquisition of the damaged properties, some property owners remained seriously interested in participating in the acquisition program. Obviously, an adequately funded program, administratively established before a storm hits, will more effectively respond to property owners who no longer wish to place themselves in a serious hazard-prone location. An established program also eliminates delays and uncertainties associated with the acquisition process and long-term site management. Such a program also serves as an appropriate link for federal funds which may be available for disaster assistance. Moreover, since many federal programs are being reduced or eliminated, it becomes essential for state programs to clearly define priorities for coastal acquisition. Finally, the Blizzard of 1978 dramatically illustrated the need to create an ongoing education program for coastal property owners and municipal and state officials to inform them of the potential benefits arising from an acquisition program for storm-damaged properties.

References

Fitzgerald, D.M., and M.O. Hayes. 1980. "Tidal Inlet Effects on Barrier Island Management." *Proceedings of Coastal Zone 1980*. Hollywood, FL: American Society of Civil Engineers, pp. 2355-2379.

Kusler, J.A. 1982. *Regulation of Flood Hazard Areas to Reduce Flood Losses*. Volume 3. Washington: U.S. Water Resources Council.

Leatherman, S.P. 1981. "Barrier Beach Development: A Perspective on the Problem." *Journal of the American Shore and Beach Protection Association* 49(2): 3-8.

U.S. Water Resources Council. 1981. *State and Local Acquisition of Floodplains and Wetlands. A Handbook on the Use of Acquisition in Floodplain Management*. Washington: Water Resources Council.

THE RETREAT ALTERNATIVE IN THE REAL WORLD: THE KILL DEVIL HILLS LAND-USE PLAN OF 1980

Raymond P. Sturza II

Town Planner
Kill Devil Hills, North Carolina

In 1972, the U.S. Congress passed the Coastal Zone Management Act (CZMA), federal legislation that was designed to preserve both the biological and economic productivity of the estuarine waters of the United States. Funding for coastal management programs authorized by CZMA would be channelled to those states that adopted their own versions of the federal mandate. The implementation technique was a simple one: States that adopted legislation establishing coastal management programs would receive federal assistance for the cost of operating such programs. States that did not adopt coastal management legislation were ineligible to receive federal assistance. As federal legislation (and funding) expired, the state programs would be established, perhaps entrenched, and coastal management would be an integral part of the budgets of each of the coastal states. Aware that their state was blessed with an abundant and highly productive system of estuarine waters and scenic beaches, the leaders of North Carolina were becoming increasingly aware of the potential for these valuable natural resources to be destroyed as the population of the so-called "frost belt" sought new playgrounds on North Carolina's undeveloped beaches. Voices of alarm rang out in the press and on the street. In 1974, the North Carolina General Assembly passed the most amended and perhaps the most controversial legislation in its history, the Coastal Area Management Act of 1974 (CAMA).

The Act created a 15-member policy commission, the Coastal Resources Commission (CRC), and charged it with the responsibility of balancing the often conflicting interests of economic development and environmental preservation. A 20-county coastal "zone" or region was delineated, using the presence of brackish water to identify which inland areas would be included in this zone under the jurisdiction of the Coastal Resources Commission. An Advisory Council composed of representatives of local governments in the region was also established in an effort to ensure that the implementation of the legislation would be a "partnership of state and local government." The legislation also established development standards and a two-tier permit system for special areas of environmental concern (AECs) and mandated the preparation of local land-use plans in the coastal region. Like the federal legislation, North Carolina's CAMA linked the availability of planning assistance funds to participation in the coastal management program. Planning assistance funds were made available to local governments in the coastal region to provide 100% of the costs of preparing local land-use plans. Unlike the federal legislation, however, nonparticipation would be penalized. At stake in North Carolina was more than the loss of planning assistance funds; communities which refused to prepared land-use plans faced the prospect of having such a document prepared, without local input, for the community by the state's Office of Coastal Management. By 1980, 19 of the region's 20 coastal counties and most municipalities had adopted local land-use plans.

Among these was the plan prepared for the Town of Kill Devil Hills (fig. 27.1). One of the elements of that plan was a prohibition against the placement of erosion abatement structures or techniques within the jurisdictional boundaries of the town. This early attempt to implement a policy of "retreat" was contained in Section One of the plan under the heading "Policies Concerning Development, 1980-1990." A total of 22 policy issues were addressed, one of which was beach and shoreline erosion. Using the results of what was described as a "random sample" public opinion survey based on the responses of 11 year-round residents and 47 non-resident property owners, the town adopted the following policy for dealing with the shoreline erosion issue:

> The Town is opposed to private and governmental actions that would attempt to stabilize the beach rather than allow it to migrate. The Town is opposed to bulkheads, jetties, groins, and the like along the oceanfront. The Town will not spend local tax monies to protect private landowners from problems due to ocean migration. The Town encourages motels to locate west of the beach road." (Coastal Consultants, Ltd., 1980).

It is interesting to note that as the winter of 1980 set in on the Town of Kill Devil Hills, not a single immovable structure was endangered due to shoreline migration. The town's policy of retreat sat silently on the shelf, hailed as progressive coastal management by the bureaucracy and the academic community. During the winter of 1982, however, the town's determination to implement this policy of retreat met its first real test.

During that winter, five subtropical storms battered the coast of northeastern North Carolina with hurricane force winds and record rainfall. The northeast orientation of the northern beaches of Dare County combined with three-foot storm tides and 18- to 24-foot waves to result in the loss of hundreds of feet of shoreline from Kitty Hawk to Oregon Inlet. Dozens of cottages were washed out to sea, septic tanks were crushed and broken, and four oceanfront motels in Kill Devil Hills were now endagered by shoreline erosion. In the spring of 1983, as the

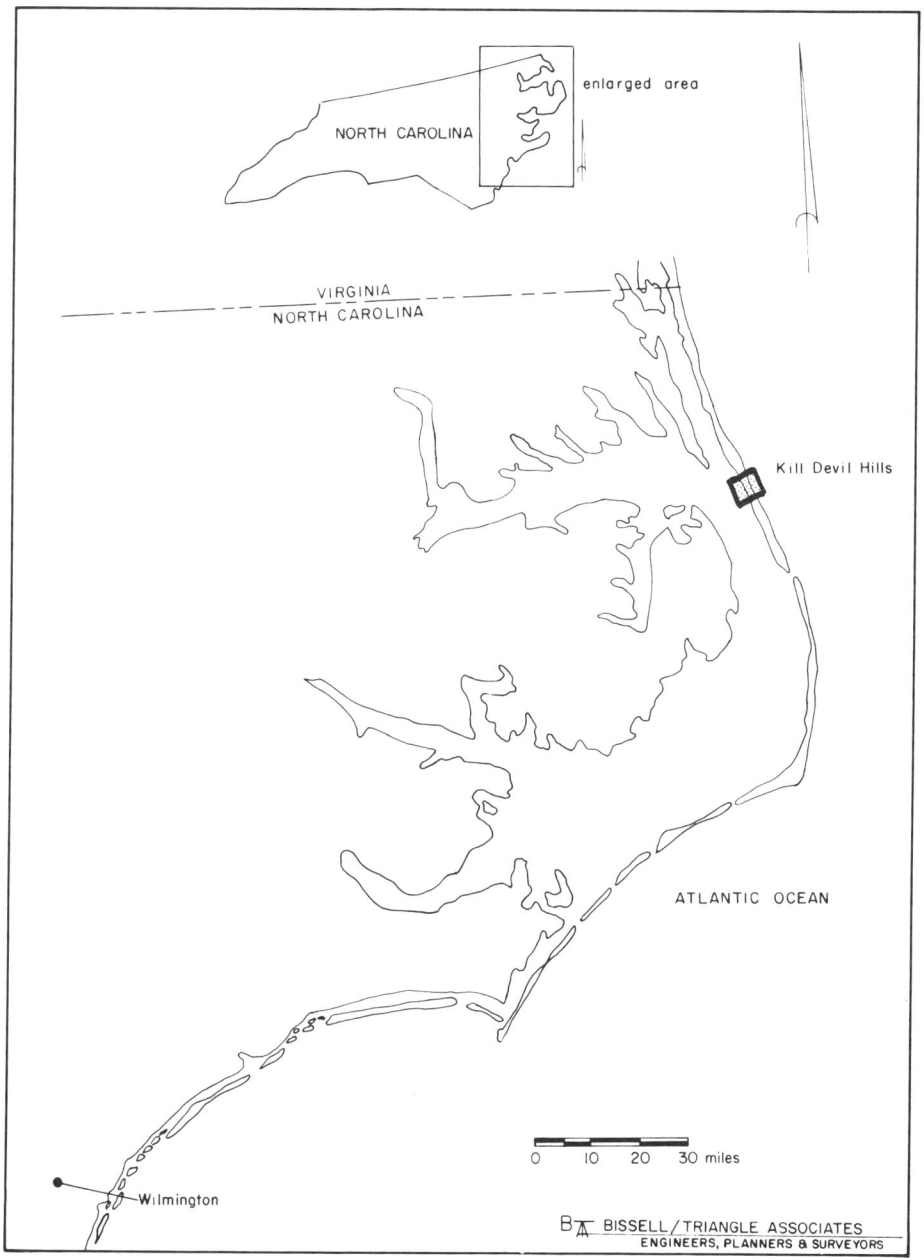

Figure 27.1. Kill Devil Hills is a coastal resort located on North Carolina's Outer Banks. Each summer the Town's year-round population of 2500 swells to around 35,000, indicating the dominance of tourism as the industrial base of the local economy. Known as the "Birthplace of Aviation", the town takes its name from the famous sand dune, Kill Devil Hill, where the Wright Brothers first piloted their "Wright Flyer" into history in 1903. Kill Devil Hill is named from historical accounts of the rum that used to wash ashore from shipwrecks laid to rest in the nearby "Graveyard of the Atlantic", Cape Hatteras, N.C. That rum, it has been said, was so potent, it could "...Kill the devil."

managers and owners of the endangered structures sought to implement erosion mitigation techniques, the Kill Devil Hills Land Use Plan emerged. Oceanfront property owners learned with disbelief that even though the State of North Carolina's standards for erosion control structures allowed endangered oceanfront structures a wide range of protection alternatives, permits for the construction of such structures could not be secured. The town's Land Use Plan included policy language opposed to public or private attempts to stabilize a migrating shoreline. As each permit application for an erosion control structure reached the North Carolina Office of Coastal Management for review, each was determined to be inconsistent with the local land-use plan and was denied. Faced with the loss of their property and/or their livelihood, the property owners responded by erecting erosion control structures, despite the denial of permits.

The plight of the oceanfront property owners and their futile battle with the sea soon caught the attention of the press. The dilapidated walkways, pools, decks, and, in some cases, living rooms of a variety of oceanfront structures in Kill Devil Hills provided a dramatic backdrop for camera crews from across the states of North Carolina and Virginia. Features on shoreline erosion and the frustrating predicament in Kill Devil Hills were run in the *Wall Street Journal, Time, Newsweek, U.S.A. Today,* and scores of state and local publications. A dichotomy of public opinion soon developed, centered around the most visible of the endangered buidings, the Sea Ranch Motel. The public was aroused on both sides of the Sea Ranch issue.

The opposing forces soon splintered into better defined groups as bayside property owners, secure in their remote setting, almost seemed to enjoy the plight of those so foolish as to locate on the oceanfront. Meanwhile, as the winter of 1983 and another storm season approached, pressure to amend the controversial land-use plan grew.

The autumn of that year brought one more wild card into the emotional setting that had already divided Kill Devil Hills into rival camps; 1983 was an election year. The terms of each of the town's five Commissioners and the Mayor would expire, and as the incumbents grappled with the shoreline erosion issue in a series of workshops with neighboring communities, a team of opponents campaigned hard on the protection issue. By November, just prior to the election, the Board of Commissioners voted to amend the town's Land Use Plan to allow the use of sandbags to protect endangered oceanfront property. All but one of the incumbents, including the mayor, were turned out of office. A new board composed of businessmen and realtors took office in December of 1983.

The winter of that year, although not as violent as the previous, brought more damage to the eroding coast. The Kill Devil Manor Motel was condemned by the local government when its septic tank broke up after heavy seas had pounded the shoreline and exposed the tank. Vandals quickly turned the deteriorating motel into the town's first oceanfront eyesore and vagrants took advantage of the site for free overnight lodging.

The policy of "retreat" had now become "abandonment" and the pressure to "do something" mounted steadily. The Planning Department received instructions to prepare an amendment to bring the town's restrictive land-use plan into consistency with the state's less restrictive standards. The Kill Devil Hills Board of

THE RETREAT ALTERNATIVE

Commissioners then met in an emergency session and passed a resolution calling for state and federal assistance in an effort to avert the further destruction of the town's shoreline.

In early 1984, as the Planning Board began the process of considering a range of alternative policies for dealing with shoreline erosion, the North Carolina Department of Natural Resources and Community Development established the Outer Banks Erosion Task Force to study the issue of erosion on the Outer Banks. The task force brought together a panel of experts with a wide range of technical and administrative skills. Several meetings were held at various locations throughout Dare County, and in July of 1984, the task force released a report that identified beach nourishment as the recommended management strategy for dealing with the erosion issue on the Outer Banks.

Following the release of that report, the Kill Devil Hills Planning Department drafted the following five alternative shoreline management policies for consideration by the Planning Board:

1. Allow the natural migration of sand and the fluctuation of the shoreline to occur unabated. Prohibit the installation of shoreline modifying structures, devices, or strategies
2. Use sandbag and soldier piling bulkheads under emergency conditions to protect endangered structures (Soldier pilings are wooden studs placed on the oceanside of sandbag bulkheads to add support in heavy seas.)
3. Allow short-term efforts to protect private or public property through the installation of bulkheads, groins, off-shore breakwaters, jetties, artificial seaweed, sandbags, revetments, rip-rap, and other similar structures or devices as may be allowed under the North Carolina Coastal Areas Management Act of 1974
4. Undertake large-scale nourishment projects to establish and maintain beaches
5. Undertake long-term efforts to protect the public beach through the funding of a beach nourishment project using state, local, and federal funds and the use of off-shore breakwaters, artificial seaweed, sandbags, relocation, acquisition and other similar measures as may be necessary as interim, temporary, remedial action to protect private or public property owners, provided such measures meet the specifications of and secure permits as required by the North Carolina Coastal Area Management Act of 1974 (CAMA).

The Planning Department staff's recommendation was in favor of alternative (5), a compromise policy that approached shoreline management from both long- and short-term perspectives. Beach nourishment was the preferred long-term response, but funding to implement a nourishment project could not be expected for at least seven years. A number of short-term responses were also recommended to cover the interim period. These measures, in order of preference, included relocation, acquisition, sandbags, off-shore breakwaters, and wooden bulkheads. Each of these measures would require littoral drift impact mitigation techniques as conditions of the permit authorizing their construction. After considerable discussion, the Planning Board agreed with the staff's recommendation and advised the Board of Commissioners of their decision. On August 20, 1984, the Kill Devil Hills Board of Commissioners held a public hearing on the proposed amendment, and an almost evenly divided audience both praised and condemned the new language. On the same day, the Board voted three to two, to adopt the policy recommended by the staff and the Planning Board. The town's brief experiment with the policy of retreat thus ended.

The Kill Devil Hills experience with the so-called "retreat alternative" exposes two of the classic pitfalls in public policy management: over-simplification of complex issues and the polarization of public opinions. On the one hand, retreat theoretically avoids the loss of beach which results from structures to protect shorefront property. On the other hand, implementation of a policy of retreat on a developed coastal barrier also ensures the inevitable loss of valuable property already firmly established on the shoreline. More often than not, those affected adversely by retreat are the community's most elite, wealthy, and influential members. Political and economic factors can come into play and the issue may become emotional, polarizing public opinion and destabilizing the governing body that is attempting to implement the policy.

Retreat from the eroding shoreline is a concept that is on sound philosophical ground. Implemented from the outset on an undeveloped coastal barrier, retreat may be the most advisable alternative to ensure the preservation of the sandy beach. However, the retroactive implementation of such a policy on an already developed barrier encounters difficulties so overwhelming that the advisability of the retreat alternative is questionable. Beach nourishment is a far more acceptable policy in the real world setting of multiple loyalties and conflicting interests.

Endnote

On February 21, 1985, the General Assembly of the Commonwealth of Virginia passed a special bill authorizing the construction of bulkheads for eight families whose homes had become endangered in the oceanfront community of Sandbridge. The special bill overruled the local Wetlands Board, which had refused to authorize permits for the construction of oceanfront bulkheads.

References

Coastal Consultants, Ltd. 1980. *Kill Devil Hills Land Use Plan.* Raleigh: North Carolina Office of Coastal Management.

TRANSFERABLE LAND RIGHTS ON DEVELOPED COASTAL BARRIERS

James T.B. Tripp

Environmental Defense Fund
New York, New York

Since coastal barrier resources provide significant ecological and economic values, applicable laws, including Section 404 of the Clean Water Act, the Flood Insurance Act, and existing state or local dune protection laws and ordinances should be vigorously enforced. Yet these resources, in particular the shallow intertidal zones and bayside wetlands, can be degraded not only by direct filling or dredging operations but by alteration of the dunes and structural efforts to restrict the natural dynamic movement of coastal barriers. This is so because seawalls and other stabilizing structures impede the natural migration of sediments and reflect wave energy. As a consequence, this may eventually lead to a deepening of the seaside and bayside shallow intertidal zones and a loss of bayside wetlands. When washover does occur, the volume of sediment moved will be less effective at maintaining barrier dimensions than would be the case under natural conditions. The removal of dune sand for fill or beach nourishment would likewise reduce the volume of material available to maintain bayside wetlands and shallow zones during migration events.

An effective, long-term program to retain the ecological values of coastal barriers requires retention or replication of the natural geological processes. In view of sea level rise, this objective necessitates the transformation to a pattern of residential use of a developed barrier which is compatible with the relentless movement of sediment shoreward and protection of natural barrier profiles. Establishment of such a pattern of residential use would be technically feasible on some developed barriers over a period of years or decades. Legally it represents a formidable challenge.

From an engineering standpoint, it is possible to build new homes so that they interfere minimally with the movement of sediment, such as by elevating them on pilings. However, while residential or commercial structures can be designed to accommodate the movement of sediment, property owners naturally desire to keep the size and configuration of their landholdings intact. They do not want to see their property transformed from apparent solid land to foredune, open beach and, eventually, to open water.

Our concepts of real property are based on fixed land masses with established boundaries. The traditional contour of our real property legal system is therefore at odds with the natural, dynamic processes of coastal barriers. While we may have the opportunity to retain natural, geological processes in public ownership, progressive development inevitably conflicts with the preservation or re-establishment of natural processes. To this end, property owners typically will import fill material, remove sediment from the dunes to fill in eroding areas, and support construction and maintenance of erosion control features.

If habitation on a developed barrier is to be consistent with natural geological processes, it must be based on a system of movable or transferable property rights which keep pace with and respond to the dynamics of the barrier. Presumably, if such a system of legal rights could be established, the social, legal, and political demand for structures and sediment-borrowing schemes intended to stabilize shorelines and private tracts would diminish. A system of movable property rights could likewise provide a sound legal basis for local, state, or federal laws which prohibit the alteration of dunes, the filling of wetlands, or the emplacement of stabilization structures on developed coastal barriers.

A system of movable property rights would have to incorporate the fact that, while a migrating barrier island loses land on the seaside, it often accretes land on the bayside. An exception may be the barrier islands off the Louisiana coast which are primarily Mississippi River deltaic fragments and therefore have a different geological origin from most other Gulf and Atlantic Coast barrier islands. Those Louisiana coastal islands may be more prone to erosion without compensating shoreward migration and land accretion as a consequence of deep Gulf water discharge of Mississippi River sediments due to the Mississippi River levees and jetties. However, since barrier islands generally are losing volume overall, a system of moveable property rights would also have to recognize the gradual shrinkage of landholdings.

The basic components of this transferable property rights system would be twofold: (1) the recognition that any new accreted bayside upland is the property of a newly established landholding entity and (2) a program of transferring development rights from eroding land on the seaside to accreting land on the bayside.

The law of ownership of accreting land is murky. However, on developed coastal barriers there is a sound public policy justification for eliminating any property claim by an adjacent landowner to bayside accreted land since that accretion depends on maintenance of natural processes. If landowners on the ocean side, to protect existing land, built erosion control structures and authorized removal of sand from higher elevations, little or no land would accrete on the bayside. In some cases, bayside land could erode as well. While restoration or retention of natural processes usually would result in bayside land accretion, most property owners would have no incentive to support technical and legal programs that promote those processes unless they have an interest in newly accreted land. Thus a legal mechanism that places ownership of land in a coastal barrier land bank is justified on sound public policy grounds.

The second component of the system is a transfer of development rights (TDR) program to transfer ownership from eroding to accreting land on a pro rata basis that would take into account changes in barrier dimensions. Few jurisdictions have much experience with TDR programs, in part because our concept of land property rights is so geometric and fixed. An extensive TDR program has recently been established in the New Jersey Pinelands.

Under the New Jersey Pinelands Commission Comprehensive Management Plan (CMP) of 1980 (pp. 210-212, 401-402), the Commission issues transferable development rights known as Pineland Development Credits (PDCs) to private landowners in the Preservation and Agricultural Production Areas where residential development is prohibited or greatly restricted. These PDCs may be sold to property owners in designated growth areas. The CMP provides for a dual-level zoning system in the growth area: a base zoning level, such as two dwelling units per acre and a bonus density level, such as three units per acre. The property owner in the growth area may build to the higher bonus density level only upon purchase of the requisite number of PDCs. The market value of PDCs therefore is dependent on land values in the designated growth areas and stems from savings on land acquisition costs.[1]

The establishment by Burlington County, the largest Pinelands county, of an Exchange to buy and sell Pineland Development Credits (TDRs) has recently been challenged in court; the New Jersey courts have sustained the program to date.[2]

As a practical matter, the effectiveness of the TDR program depends on one institution controlling use or ownership of newly accreted land and issuance of the transferable land rights. The land bank which holds title to newly accreted lands would be authorized to issue rights to oceanfront property owners. Those rights would gradually mature as individual oceanfront tracts erode. Upon maturity, the seaside property owner would then use that right to acquire title to newly accreted upland from the bank. The bank would have to determine the relationship between the size of lots on the newly accreted upland and the seaside eroded lands as a function of overall island volume. Rights turned into the bank in exchange for new bayside lots fringing on accreting bayside wetlands would then be extinguished.

This new system has some distinct advantages over our traditional system of land ownership; it also has certain drawbacks. The overall advantage of this movable property rights program is that it is derived from and therefore compatible with the geologic nature of the land resource itself. It is a system that could be considered only if the requisite local, state, or federal decision makers and affected property owners decided to retain or, over time, reinstate natural processes. Under our present system, developed barrier property owners are faced with an all-or-nothing choice: either they fight natural processes with private or public works at increasing economic and environmental costs or they watch their landholdings eventually wither away.

---

[1] For a general discussion of TDR programs, see J.J. Costonis, 1975, "Development Rights Transfer: An Exploratory Essay," 83 Yale L.J. 75.
[2] See *Matlack v. Burlington County Board of Chosen Freeholders*, 191 N.J. Super. 136 (Law Div. 1982), *aff'd. per curiam*. DKt. No. A-6028-82T3 (App. Div. June 14, 1984).

This movable property system provides an alternative—retention of a property right, i.e., a buildable lot on the coastal barrier, but relatively little control over the precise location, size, or configuration of that lot. Since it would be impractical to move all tracts at once, this system moves the oceanfront property owners gradually to the bayside. In turn, a new set of oceanfront property owners then appears who would be next in line to benefit from the island's landholding bank TDR program. Thus, while the program would decidedly not guarantee any property owner a permanent seaside tract, it would guarantee any property owner a lot somewhere on the barrier. To many property owners, this feature may constitute a serious disadvantage to the program.

If maintenance of barrier islands, however, is not publicly subsidized and local, state, and federal law protects key barrier resources, this property system may have more appeal to private property owners than other alternatives. If the relocation of oceanfront property owners to the bayside is viewed as a particularly objectionable feature of a movable property system, a variation would be to have all of the property owners move one house toward the bayside when the most oceanside lot erodes.

This property rights system will only accomplish its resource objective if the island bank controls the design of residential or commercial structures. In particular, only those designs that are compatible with natural geological processes would be allowed, so as not to interfere with sediment transport.

Residents of developed coastal barriers have differing levels of commitment to preserving a fixed location in opposition to natural processes. While certain landowners may decide to pursue the structural approach to land stability to any extreme that is feasible from an engineering perspective, at a minimum these landowners should be compelled to internalize the real economic costs of that decision. Public subsidies for such activities should cease altogether.

For any developed coastal barrier, however, a transformation to the new property system is theoretically feasible. That transition necessitates the cessation of new investment in structural, erosion-control works. Further, when storms cause substantial damage to existing residential or commercial structures, authorization for reconstruction in place should require compliance with design specifications that are compatible with natural processes. During the transformation process, oceanside landowners must have confidence that they will obtain title to new land elsewhere on the barrier.

Since the external costs of shoreline stabilization programs are substantial and rising, the decision to make the transition to the new movable property rights system should not and cannot rest with the present generation of property owners alone. The legal system described here provides planners and property owners with a conceptual alternative which promotes continued private residential and recreational use of a developed coastal barrier in a manner compatible with natural geological processes. For planners and public officials, a well-designed legal system, establishing a central landholding bank or comparable institution, and a TDR program should provide a sound basis for meeting anticipated legal challenges.

APPENDICES

COASTAL BARRIER RESOURCES ACT OF 1982
P.L. 97-348; 16 U.S.C. 3501-10

Ninety-seventh Congress of the United States of America

AT THE SECOND SESSION

Begun and held at the City of Washington on Monday, the twenty-fifth day of January, one thousand nine hundred and eighty-two

An Act

To protect and conserve fish and wildlife resources, and for other purposes.

Be it enacted by the Senate and House of Representatives of the United States of America in Congress assembled.

SECTION 1. SHORT TITLE.

This Act may be cited as the "Coastal Barrier Resources Act".

SEC. 2. FINDINGS AND PURPOSE.

(a) FINDINGS.—The Congress finds that—

(1) coastal barriers along the Atlantic and Gulf coasts of the United States and the adjacent wetlands, marshes, estuaries, inlets and nearshore waters provide—

(A) habitats for migratory birds and other wildlife; and

(B) habitats which are essential spawning, nursery, nesting, and feeding areas for commercially and recreationally important species of finfish and shellfish, as well as other aquatic organisms such as sea turtles;

(2) coastal barriers contain resources of extraordinary scenic, scientific, recreational, natural, historic, archeological, cultural, and economic importance; which are being irretrievably damaged and lost due to development on, among, and adjacent to, such barriers;

(3) coastal barriers serve as natural storm protective buffers and are generally unsuitable for development because they are vulnerable to hurricane and other storm damage and because natural shoreline recession and the movement of unstable sediments undermine manmade structures;

(4) certain actions and programs of the Federal Government have subsidized and permitted development on coastal barriers and the result has been the loss of barrier resources, threats to human life, health, and property, and the expenditure of millions of tax dollars each year; and

(5) a program of coordinated action by Federal, State, and local governments is critical to the more appropriate use and conservation of coastal barriers.

(b) PURPOSE.—The Congress declares that it is the purpose of this Act to minimize the loss of human life, wasteful expenditure of Federal revenues, and the damage to fish, wildlife, and other natural resources associated with the coastal barriers along the Atlantic and Gulf coasts by restricting future Federal expenditures and financial assistance which have the effect of encouraging development of coastal barriers, by establishing a Coastal Barrier Resources System, and by considering the means and measures by which the long-term conservation of these fish, wildlife, and other natural resources may be achieved.

SEC. 3. DEFINITIONS.

For purposes of this Act—

S. 1018—2

(1) The term "undeveloped coastal barrier" means—
　(A) a depositional geologic feature (such as a bay barrier, tombolo, barrier spit, or barrier island) that—
　　(i) consists of unconsolidated sedimentary materials,
　　(ii) is subject to wave, tidal, and wind energies, and
　　(iii) protects landward aquatic habitats from direct wave attack; and
　(B) all associated aquatic habitats, including the adjacent wetlands, marshes, estuaries, inlets, and nearshore waters;
but only if such feature and associated habitats (i) contain few manmade structures and these structures, and man's activities on such feature and within such habitats, do not significantly impede geomorphic and ecological processes, and (ii) are not included within the boundaries of an area established under Federal, State, or local law, or held by a qualified organization as defined in section 170(h)(3) of the Internal Revenue Code of 1954, primarily for wildlife refuge, sanctuary, recreational, or natural resource conservation purposes.

(2) The term "Committees" refers to the Committee on Merchant Marine and Fisheries of the House of Representatives and the Committee on Environment and Public Works of the Senate.

(3) The term "financial assistance" means any form of loan, grant, guaranty, insurance, payment, rebate, subsidy, or any other form of direct or indirect Federal assistance other than—
　(A) general revenue-sharing grants made under section 102 of the State and Local Fiscal Assistance Amendments of 1972 (31 U.S.C. 1221);
　(B) deposit or account insurance for customers of banks, savings and loan associations, credit unions, or similar institutions;
　(C) the purchase of mortgages or loans by the Government National Mortgage Association, the Federal National Mortgage Association, or the Federal Home Loan Mortgage Corporation;
　(D) assistance for environmental studies, planning, and assessments that are required incident to the issuance of permits or other authorizations under Federal law; and
　(E) assistance pursuant to programs entirely unrelated to development, such as any Federal or federally assisted public assistance program or any Federal old-age survivors or disability insurance program.
Effective October 1, 1983, such term includes flood insurance described in section 1321 of the National Flood Insurance Act of 1968, as amended (42 U.S.C. 4028).

(4) The term "Secretary" means the Secretary of the Interior.
(5) The term "System unit" means any undeveloped coastal barrier, or combination of closely-related undeveloped coastal barriers, included within the Coastal Barrier Resources System established by section 4.

SEC. 4. THE COASTAL BARRIER RESOURCES SYSTEM.

(a) ESTABLISHMENT.—(1) There is established the Coastal Barrier Resources System which shall consist of those undeveloped coastal barriers located on the Atlantic and Gulf coasts of the United States that are identified and generally depicted on the maps that are

S. 1018—3

entitled "Coastal Barrier Resources System", numbered A01 through T12, and dated September 30, 1982.

(2) Any person or persons or other entity owning or controlling land on an undeveloped coastal barrier, associated landform or any portion thereof not within the Coastal Barrier Resources System established under paragraph (1) may, within one year after the date of enactment of this Act, elect to have such land included within the Coastal Barrier Resources System. This election shall be made in compliance with regulations established for this purpose by the Secretary not later than one hundred and eighty days after the date of enactment of this Act; and, once made and filed in accordance with the laws regulating the sale or other transfer of land or other real property of the State in which such land is located, shall have the same force and effect as if such land had originally been included within the Coastal Barrier Resources System.

(b)(1) As soon as practicable after the enactment of this Act, the maps referred to in paragraph (1) of subsection (a) shall be filed with the Committees by the Secretary, and each such map shall have the same force and effect as if included in this Act, except that correction of clerical and typographical errors in each such map may be made. Each such map shall be on file and available for public inspection in the Office of the Director of the United States Fish and Wildlife Service, Department of the Interior, and in other appropriate offices of the Service.

(2) As soon as practicable after the date of the enactment of this Act, the Secretary shall provide copies of the maps referred to in paragraph (1) of subsection (a) to the chief executive officer of (A) each State and county or equivalent jurisdiction in which a system unit is located, (B) each State coastal zone management agency in those States which have a coastal zone management plan approved pursuant to section 306 of the Coastal Zone Management Act of 1972 (16 U.S.C. 1455) and in which a system unit is located, and (C) each appropriate Federal agency.

(c) BOUNDARY MODIFICATIONS.—(1) Within 180 days after the date of enactment of this Act, the Secretary may make such minor and technical modifications to the boundaries of system units as depicted on the maps referred to in paragraph (1) of subsection (a) as are consistent with the purposes of this Act and necessary to clarify the boundaries of said system units; except that, for system units within States which have, on the date of enactment, a coastal zone management plan approved pursuant to section 306 of the Coastal Zone Management Act of 1972 (16 U.S.C. 1455)—

(A) each appropriate State coastal zone management agency may, within 90 days after the date of enactment of this Act, submit to the Secretary proposals for such minor and technical modifications; and

(B) the Secretary may, within 180 days after the date of enactment of this Act, make such minor and technical modifications to the boundaries of such system units.

(2) The Secretary shall, not less than 30 days prior to the effective date of any such boundary modification made under the authority of paragraph (1), submit written notice of such modification to (A) each of the Committees and (B) each of the appropriate officers referred to in paragraph (2) of subsection (b).

(3) The Secretary shall conduct, at least once every five years, a review of the maps referred to in paragraph (1) of subsection (a) and make, in consultation with the appropriate officers referred to in

S. 1018—4

paragraph (2) of subsection (b), such minor and technical modifications to the boundaries of system units as are necessary solely to reflect changes that have occurred in the size or location of any system units as a result of natural forces.

(4) If, in the case of any minor and technical modification to the boundaries of system units made under the authority of this subsection, an appropriate chief executive officer of a State, county or equivalent jurisdiction, or State coastal zone management agency to which notice was given in accordance with this subsection files comments disagreeing with all or part of the modification and the Secretary makes a modification which is in conflict with such comments, or if the Secretary fails to adopt a modification pursuant to a proposal submitted by an appropriate State coastal zone management agency under paragraph (1)(A), the Secretary shall submit to the chief executive officer a written justification for his failure to make modifications consistent with such comments or proposals.

SEC. 5. LIMITATIONS ON FEDERAL EXPENDITURES AFFECTING THE SYSTEM.

(a) Except as provided in section 6, no new expenditures or new financial assistance may be made available under authority of any Federal law for any purpose within the Coastal Barrier Resources System, including, but not limited to—

(1) the construction or purchase of any structure, appurtenance, facility, or related infrastructure;

(2) the construction or purchase of any road, airport, boat landing facility, or other facility on, or bridge or causeway to, any System unit; and

(3) the carrying out of any project to prevent the erosion of, or to otherwise stabilize, any inlet, shoreline, or inshore area, except that such assistance and expenditures may be made available on units designated pursuant to section 4 on maps numbered S01 through S08 for purposes other than encouraging development and, in all units, in cases where an emergency threatens life, land, and property immediately adjacent to that unit.

(b) An expenditure or financial assistance made available under authority of Federal law shall, for purposes of this Act, be a new expenditure or new financial assistance if—

(1) in any case with respect to which specific appropriations are required, no money for construction or purchase purposes was appropriated before the date of the enactment of this Act; or

(2) no legally binding commitment for the expenditure or financial assistance was made before such date of enactment.

SEC. 6. EXCEPTIONS.

(a) Notwithstanding section 5, the appropriate Federal officer, after consultation with the Secretary, may make Federal expenditures or financial assistance available within the Coastal Barrier Resources System for—

(1) any use or facility necessary for the exploration, extraction, or transportation of energy resources which can be carried out only on, in, or adjacent to coastal water areas because the use or facility requires access to the coastal water body;

S. 1018—5

(2) the maintenance of existing channel improvements and related structures, such as jetties, and including the disposal of dredge materials related to such improvements;

(3) the maintenance, replacement, reconstruction, or repair, but not the expansion, of publicly-owned or publicly-operated roads, structures, or facilities that are essential links in a larger network or system;

(4) military activities essential to national security;

(5) the construction, operation, maintenance, and rehabilitation of Coast Guard facilities and access thereto; and

(6) any of the following actions or projects, but only if the making available of expenditures or assistance therefor is consistent with the purposes of this Act:

(A) Projects for the study, management, protection and enhancement of fish and wildlife resources and habitats, including, but not limited to, acquisition of fish and wildlife habitats and related lands, stabilization projects for fish and wildlife habitats, and recreational projects.

(B) The establishment, operation, and maintenance of air and water navigation aids and devices, and for access thereto.

(C) Projects under the Land and Water Conservation Fund Act of 1965 (16 U.S.C. 460l-4 through 11) and the Coastal Zone Management Act of 1972 (16 U.S.C. 1451 et seq.).

(D) Scientific research, including but not limited to aeronautical, atmospheric, space, geologic, marine, fish and wildlife and other research, development, and applications.

(E) Assistance for emergency actions essential to the saving of lives and the protection of property and the public health and safety, if such actions are performed pursuant to sections 305 and 306 of the Disaster Relief Act of 1974 (42 U.S.C. 5145 and 5146) and section 1362 of the National Flood Insurance Act of 1968 (42 U.S.C. 4103) and are limited to actions that are necessary to alleviate the emergency.

(F) The maintenance, replacement, reconstruction, or repair, but not the expansion, of publicly owned or publicly operated roads, structures, or facilities.

(G) Nonstructural projects for shoreline stabilization that are designed to mimic, enhance, or restore natural stabilization systems.

(b) For purposes of subsection (a)(2), a channel improvement or a related structure shall be treated as an existing improvement or an existing related structure only if all, or a portion, of the moneys for such improvement or structure was appropriated before the date of the enactment of this Act.

SEC. 7. CERTIFICATION OF COMPLIANCE.

The Director of the Office of Management and Budget shall, on behalf of each Federal agency concerned, make written certification that each such agency has complied with the provisions of this Act during each fiscal year beginning after September 30, 1982. Such certification shall be submitted on an annual basis to the House of Representatives and the Senate pursuant to the schedule required under the Congressional Budget and Impoundment Control Act of 1974.

S. 1018—6

SEC. 8. PRIORITY OF LAWS.

Nothing contained in this Act shall be construed as indicating an intent on the part of the Congress to change the existing relationship of other Federal laws to the law of a State, or a political subdivision of a State, or to relieve any person of any obligation imposed by any law of any State, or political subdivision of a State. No provision of this Act shall be construed to invalidate any provision of State or local law unless there is a direct conflict between such provision and the law of the State, or political subdivision of the State, so that the two cannot be reconciled or consistently stand together. This Act shall in no way be interpreted to interfere with a State's right to protect, rehabilitate, preserve, and restore lands within its established boundary.

SEC. 9. SEPARABILITY.

If any provision of this Act or the application thereof to any person or circumstance is held invalid, the remainder of the Act and the application of such provision to other persons not similarly situated or to other circumstances shall not be affected thereby.

SEC. 10. REPORTS TO CONGRESS.

(a) IN GENERAL.—Before the close of the 3-year period beginning on the date of the enactment of this Act, the Secretary shall prepare and submit to the Committees a report regarding the System.

(b) CONSULTATION IN PREPARING REPORT.—The Secretary shall prepare the report required under subsection (a) in consultation with the Governors of the States in which System units are located and with the coastal zone management agencies of the States in which System units are located and after providing opportunity for, and considering, public comment.

(c) REPORT CONTENT.—The report required under subsection (a) shall contain—

(1) recommendations for the conservation of the fish, wildlife, and other natural resources of the System based on an evaluation and comparison of all management alternatives, and combinations thereof, such as State and local actions (including management plans approved under the Coastal Zone Management Act of 1972 (16 U.S.C. 1451 et seq.)), Federal actions (including acquisition for administration as part of the National Wildlife Refuge System), and initiatives by private organizations and individuals;

(2) recommendations for additions to, or deletions from, the Coastal Barrier Resources System, and for modifications to the boundaries of System units;

(3) a summary of the comments received from the Governors of the States, State coastal zone management agencies, other government officials, and the public regarding the System; and

(4) an analysis of the effect, if any, that general revenue sharing grants made under section 102 of the State and Local Fiscal Assistance Amendments of 1972 (31 U.S.C. 1221) have had on undeveloped coastal barriers.

SEC. 11. AMENDMENTS REGARDING FLOOD INSURANCE.

(a) Section 1321 of the National Flood Insurance Act of 1968 (42 U.S.C. 4028) is amended to read as follows:

S. 1018—7

"UNDEVELOPED COASTAL BARRIERS

"SEC. 1321. No new flood insurance coverage may be provided under this title on or after October 1, 1983, for any new construction or substantial improvements of structures located on any coastal barrier within the Coastal Barrier Resources System established by section 4 of the Coastal Barrier Resources Act. A federally insured financial institution may make loans secured by structures which are not eligible for flood insurance by reason of this section.".

(b) Section 341(d)(2) of the Omnibus Budget and Reconciliation Act of 1981 (Public Law 97-35) is repealed.

SEC. 12. AUTHORIZATION OF APPROPRIATIONS.

There is authorized to be appropriated to the Department of the Interior $1,000,000 for the period beginning October 1, 1982, and ending September 30, 1985, for purposes of carrying out sections 4 and 10.

Speaker of the House of Representatives.

Vice President of the United States and President of the Senate.

Biographical Sketches of the Authors

EARL J. "JAY" BAKER is an Associate Professor in the Department of Geography and Director of the Environmental Hazards Center at Florida State University. He is also President of Hazards Management Group, Inc. Since 1971, he has been active in research concerning individual and institutional response to environmental hazards.

TIMOTHY BEATLEY is a Research Associate with the Center for Urban and Regional Studies at the University of North Carolina at Chapel Hill. He has taught planning theory and land-use policy and has for the last two years been involved in an NSF-funded research project examining the role of development management in reducing damages from hurricanes and coastal storms. He holds an M.A. in Political Science from UNC-Chapel Hill, an M.U.P. from the University of Oregon, and a B.C.P. from the University of Virginia.

MARK A. BENEDICT received his Bachelor of Science degree from Duke University (1974), and his Master of Science and Doctorate from the University of Massachusetts at Amherst (1978, 1981). During studies at the Duke University Marine Laboratory, research with the University of Massachusetts National Park Service Cooperative Research Unit, and a one-year appointment at the University of Florida, he investigated many Atlantic coastal barriers and worked on programs for their management. At the time of the "Cities on the Beach" conference, he was Director of the Natural Resources Management Department for Collier County, Florida. Mark is now Director of Environmental Protection for The Conservancy, Inc. of Naples, Florida.

DAVID J. BROWER, a planner and lawyer, is Associate Director of the Center for Urban and Regional Studies at the University of North Carolina at Chapel Hill. He is active in the areas of growth management, coastal zone management, and natural hazards. He holds B.A. and J.D. degrees from the University of Michigan.

BARBARA K. R. BURBANK is a coastal geographer. She received her B.A. and M.S. from the University of Massachusetts Department of Geology and Geography. She was Assistant Coordinator of the 1985 Conference on the Management of Developed Coastal Barriers. Research interests include hazard perception, coastal barrier management, and education. Barbara currently resides on Cape Cod.

KEANE CALLAHAN received a Bachelor of Science Degree in Wildlife Biology in 1982 from Colorado State University. He recently completed his Master's Degree in Regional Planning at the University of Massachusetts, Amherst. His thesis research assessed the immediate and long-term impacts on developed coastal barriers which result from fragmentation of political authority. He is presently working as an environmental planner with the King's Mark Environmental Review Team in Wallingford, Connecticut.

ALAN P. CHESNEY, Ph.D., is the coordinator of the Employee Assistance Program at the University of Texas Medical Branch in Galveston. He currently holds faculty appointments in the Department of Psychiatry and Behavioral Sciences and the Department of Preventive Medicine and Community Health. Dr. Chesney's recent publications are in the areas of evaluation of mental health service delivery and the impact of anger expression on high blood pressure.

GARY R. CLAYTON is a graduate of Rutgers University and the University of Massachusetts, Amherst. While presenting his paper at the "Cities on the Beach" conference, he was Assistant Director of the Massachusetts Coastal Zone Management Office. With the CZM program, he helped develop and implement public policy for the waterfront development, critical areas management, coastal natural hazards, and water resources management. Gary is now Director for the Division of Wetlands and Waterways Regulation within the Massachusetts Department of Environmental Quality Engineering.

ALEXANDRA D. DAWSON is Co-Director of the Water Supply Citizens Advisory Committee, a state-funded citizens' group watchdogging Massachusetts waters supply policies. She directs a graduate program in Resource Management and Administration at Antioch University's New England Division and teaches at the University of Massachusetts. She is President-Elect of the Massachusetts Association of Conservation Commissions; she is author of *Land Use Planning and the Law* (Garland, NY, 1982).

CHRISTOPHER J. DUERKSEN is a Senior Associate with the Conservation Foundation and Adjunct Professor of Historic Preservation Law and Economics at Mary Washington College. He is currently a Co-Project Director of the Conservation Foundation's Use of Land Project, a comprehensive study of land use and growth in the United States. Mr. Duerksen directed research for the Foundation's

Industrial Plant Siting Program and conducted a project on low-level radioactive waste management. He is the co-author/editor of several publications, including a recent book on the future of the National Parks and books on river conservation (1984) and historic preservation law (1983). Before joining the Conservation Foundation, Mr. Duerksen worked with the Chicago Law firm of Ross, Hardies, O'Keefe, Babcock and Parsons on environmental projects. Mr. Duerksen received a law degree from the University of Chicago in 1974.

OWEN J. FURUSETH is an Associate Professor of Geography at the University of North Carolina at Charlotte. His teaching and research interests are in the area of land-use planning and management. Prior to joining academia, he was a planner with the Jacksonville (Florida) area Planning Board and was involved in coastal area planning.

PAUL A. GARES was Assistant Professor of Geography at Ohio University when the "Cities on the Beach" conference was held. He holds a Ph.D. from Rutgers University. His dissertation deals with differences in dune formation processes along undeveloped and developed shorelines. He has also participated in a variety of coastal geomorphology research projects funded by the National Park Service. Paul is now Assistant Professor of Geography at Colgate University.

PAUL JEFFREY GODFREY received his Ph.D. from Duke University in 1969 and has taught in the Botany Department of the University of Massachusetts at Amherst since 1970. His research interests include coastal plant ecology, plant geography, aquatic plants, and the impacts of recreation on coastal environments. He has undertaken much research for the National Park Service and is a past Director of the NPS. Professor Godfrey is a former Chairperson and active member of the area's Five-College Marine Science Program. He and his wife Melinda, a marine biologist, have collaborated on many research projects and have published extensively.

DAVID R. GODSCHALK is Professor of City and Regional Planning at the University of North Carolina at Chapel Hill. He is author of *Impacts of the Coastal Barrier Resources Act: A Pilot Study* (Washington, DC: Office of Ocean and Coastal Resources Management, 1984); co-author of *Before the Storm: Managing Development to Reduce Hurricane Damages* (Chapel Hill, NC: Center for Urban and Regional Studies, 1982); and co-investigator on an NSF-funded national study of the use of development management to mitigate hurricane damage.

RICHARD HAMANN is Acting Director of the Center for Governmental Responsibility at the University of Florida in Gainesville. He received his J.D. in 1976 from the University of Florida College of Law. He worked at the Eastern Water

Law Center until transferring in 1980 to the Center for Governmental Responsibility. He served as Assistant Director of the Center before assuming his present role. Professor Hamann teaches courses in Land Use, Environmental Law, and Water Law. He is currently conducting a study on post-hurricane reconstruction planning for the Florida Sea Grant College.

STEPHEN N. HUFFMAN, Galveston City Manager, attended Texas Christian University, where he majored in Public Administration with a minor in Economics. Mr. Huffman worked for the City of Galveston during summers while attending college. In 1972, he returned to work for the City of Galveston and has held several key positions, first as Administrative Assistant to the City Manager and Public Safety Director.

SALLIE M. IVES , an Associate Professor of Geography at the University of North Carolina at Charlotte, received a B.A. and M.A. from the University of Maryland at College Park and a Ph.D. from the University of Illinois at Urbana. Her current research and teaching interests are in the areas of social impact assessment and applied population analysis. She is the author or co-author of several articles in such journals as *Environment and Behavior* and the *Southeastern Geographer*, monographs, numerous research reports, and articles in housing trade journals. Her current research on community response to beach erosion hazard is funded partially by the National Hazards Research and Application Information Center in Boulder, Colorado. She is currently working on a beach erosion monograph with Dr. Owen J. Furuseth, also of UNCC.

ROBERT JANISKEE is an Associate Professor of Geography and Geography Internship Program Director at the University of South Carolina. His principal teaching interests include recreation geography and environmental resources management, and his research specialties are family camping, festivals, and rural recreation in the South.

CHRISTOPHER P. JONES received his Master's Degree in Coastal and Oceanographic Engineering from the University of Florida in 1977. After working with a private coastal engineering consultant for two years, he joined the Florida Sea Grant Extension Program as their coastal engineering specialist. In 1986, he joined Coastal Science & Engineering, Inc., in Columbia, South Carolina. His areas of interest include hydraulics and stability of tidal inlets, erosion control, dune restoration, marinas, and hurricane resistant construction.

STEPHEN P. LEATHERMAN is a geomorphologist with the Department of Geography at the University of Maryland. Dr. Leatherman has previously held academic positions at Boston University, the University of Massachusetts at Amherst, and Yale University. He also served as Director of National Park Service Cooperative Research Unit at the University of Massachusetts before going to Maryland. His research interests involve barrier island dynamics.

PAUL LOVINGOOD is Professor of Geography at the University of South Carolina. His research interests include computer mapping, population migration, and the patterns and structure of recreation.

MICHAEL MANTELL is an attorney and a Senior Associate with The Conservation Foundation in Washington, DC. He is co-author of *National Parks for a New Generation: Visions, Realities, Prospects*, has contributed to the State of the Environment reports, and has been involved in a variety of land-use matters for The Conservation Foundation, including coastal zone management and historic preservation. He is chairman of an American Bar Association Subcommittee on Federal Land Use Policy.

JAMES M. McCLOY is Professor of Marine Geography, Director of the Coastal Zone Laboratory, and Director of the Center for Marine Training and Safety at the Texas A & M University at Galveston. His publications include "Water-Related Fatalities—The Texas Study" (1980), "Guidelines for Establishing Open-Water Recreational Beach Standards" (1981), "Planning Protective and Rescue Services at Open-Water Recreational Beaches Using Epidemiological Data" (1982), "Analysis of the Population-at-Risk in Regard to Water-Related Activities in Coastal Texas" (1983), "Alcohol and Drownings—Facts and Folk Truths" (1984), and "Man's Response to Coastal Change in the Northern Gulf of Mexico," with D. Davis and A. Craig (in press).

H. CRANE MILLER is an attorney at law in Washington, D.C. He received an A.B. from Williams College in 1957 and an L.L.B. from the University of Virginia Law School in 1960. Since then he has performed research and written extensively on coastal development issues, including the prevention and mitigation of natural hazards in developed coastal areas.

J. KENNETH MITCHELL is Professor of Geography and Director of the Graduate Program in Geography at Rutgers University. He received degrees from Queens University, Belfast (B.Sc., 1965), University of Cincinnati (M.A. 1967. M.C.P., 1967), and the University of Chicago (Ph.D., 1973). He has served as Chairman of the U.S. Scientific Committee on the Outer Continental Shelf (1979-81), the Committee on Natural Hazard Vulnerability and Hazard Mitigation (National Research Council, 1983), and member of the Committee on Natural Disasters (National Research Council, 1982-present).

BIOGRAPHICAL SKETCHES

HANS N. NEUHAUSER has served as the Coastal Director of the Georgia Conservancy in Savannah since 1972. He is a board member of the Coastal Area Planning and Development Commission, and Chairman of its Resource Management and Long-Range Planning Committee, among numerous other planning involvements.

KARL F. NORDSTROM is Associate Research Professor with the Center for Coastal and Environmental Studies at Rutgers University. His current research has been directed toward an understanding of dune dynamics and the behavior of the ocean, bay, and tidal inlet beaches which comprise the barrier island system. Research has also been directed toward an analysis of natural hazards and land use in the coastal zone.

SHEILA G. PELCZARSKI is a Master's Degree candidate in the Department of Geology and Geography at the University of Massachusetts at Amherst, where she is studying the use of satellite remote sensing for natural resource management. She is also staff assistant with the Land and Water Policy Center. Sheila has extensive experience with coastal research and management issues and has edited numerous articles and manuscripts in the coastal field. She was also involved in the original data gathering effort which resulted in the inventory now known as the Coastal Barrier Resources System. This volume represents her second major co-editing effort, following the publication of *Ecological Considerations for Wetlands Treatment of Municipal Wastewaters* by Van Nostrand Reinhold in 1985.

ORRIN H. PILKEY is James B. Duke Professor of Geology at Duke University. His research specialties are in continental margin sedimentation and coastal geology. He is particularly concerned with geological/oceanographic processes on developed barrier islands. Currently he is co-editing a 23-volume series of layperson-level, coastal hazard books for each of the coastal states and two Great Lakes.

RUTHERFORD H. PLATT is Professor of Geography and Planning Law and Director of the Land and Water Policy Center at the University of Massachusetts, Amherst. He graduated from Yale University with a B.A. in Political Science in 1962. He holds a law degree (J.D.) and Ph.D. in Geography, both from the University of Chicago, and he is a member of the Illinois Bar. Dr. Platt specializes in problems relating to the management of land and water resources in the U.S. He has worked as researcher and consultant in the areas of urban open space preservation, farmland retention, urban renewal, and land-use zoning. His major focus over the past five years has been the management of floodplains, wetlands, and coastal areas.

BIOGRAPHICAL SKETCHES

NORBERT P. PSUTY is Director of the Center for Coastal and Environmental Studies, a research unit of Rutgers University. He holds a Ph.D. in Geography from Louisiana State University where he was a Research Associate with the Coastal Studies Institute. He has conducted research on coastal geomorphology in Latin America, concentrating on beach-ridge development and sea level changes. Further research has been directed toward dune-beach interaction and dune mobility in coastal New Jersey and New York. Dr. Psuty has applied his geomorphological background in barrier island dynamics by participating as a member of the Barrier Island Task Force of the Department of the Interior. His current research emphasis is on barrier island mobility and coastal dunes.

JONATHAN SILBERMAN is Professor of Economics and Chairman of the Department of Economics and Finance at the University of·Baltimore. His primary research area is applied microeconomics. He recently completed a cost/benefit evaluation of the proposed Atlantic Avenue Beautification Plan for the City of Virginia Beach. Recent publications include "Forecasting Recreation Demand: An Application of the Travel Cost Model" (*Review of Regional Studies*), "A Demand Function for Length of Stay: The Evidence from Virginia Beach" (*Journal of Travel Research*), and "Differential Response to Change: The Case of Home Purchase" (*Journal of Urban Economics*).

BYRON D. SPANGLER is Professor Emeritus of Civil Engineering at the University of Florida. In addition to thirty-seven years of teaching there, he has taught numerous courses on fallout shelter analysis and design, and multi-disaster design. He has investigated hurricane damage to structures and written several papers on the topic. He was the principal investigator on a study of hurricane shelters in the Florida Keys.

RAYMOND P. STURZA II is Town Planner for the Town of Kill Devil Hills, North Carolina. Mr. Sturza received a B.A. degree in Political Science from the University of North Carolina at Wilmington and an M.S. degree in Public Administration from East Carolina University.

JAMES T. B. TRIPP has acted as counsel for the Environmental Defense Fund since 1973. He graduated from Yale Law School with an L.L.B. in 1966. He was Assistant U.S. Attorney for the Southern District of New York from 1968 to 1973.

NEILS WEST received a Ph.D. in Geography from Rutgers University and has long been interested in a wide range of coastal resource management issues, including coastal demographics. He is currently associated with the Department of Geography and Marine Affairs at the University of Rhode Island. In addition to teaching courses in Coastal Management, Environmental Management, and Impact Assessment, he is also Co-Director of the Landsat Remote Sensing Lab.

Selected Bibliography on Coastal Barriers

Compiled by Rutherford H. Platt,

Barbara K. R. Burbank,

Jean Slosek, and Sheila Pelczarski

BOOKS

Barth, M.C., and J.G. Titus. 1984. *Greenhouse Effect and Sea Level Rise.* New York: Van Nostrand Reinhold.

Brinkmann, W.A.R., et al. 1975. *Hurricane Hazard in the United States: A Research Assessment.* Environment and Man Monograph no. 007. Boulder: University of Colorado, Institute of Behavioral Science.

Burton, I., R.W. Kates, and R.E. Snead. 1969. *The Human Ecology of Coastal Flood Hazard in Megalopolis.* Research Paper no. 115. Chicago: The University of Chicago, Department of Geography.

Canis, W.F., et al. 1984. *Living with the Alabama-Mississippi Shore.* Durham: Duke University Press.

Ditton, R.B., J.L. Seymour, and G.C. Swanson. 1977. *Coastal Resources Management.* Lexington: Lexington Press.

Doyle, L.J., et al. *Living with the West Florida Shore.* Durham: Duke University Press.

Ducsik, D.W. 1974. *Shoreline for the Public: A Handbook of Social and Legal Considerations regarding Public Recreational Use of the Nation's Coastal Shoreline.* Cambridge: M.I.T. Press.

Ford, J.C., A.K. Horme, and E.J. Pullen. 1983. *An Annotated Bibliography on the Biological Effects of Constructing Channels, Jetties, and Other Coastal Structures.* Fort Belvoir: Coastal Engineering Research Center, U.S. Army Corps of Engineers.

Gares, P.A., et al. 1979. *Coastal Dunes: Their Function, Delineation and Management.* New Brunswick: Rutgers University, Center for Coastal and Environmental Studies.

Heikoff, J.M. 1976. *The Politics of Shore Erosion: Westhampton Beach.* Ann Arbor: Ann Arbor Science.

_____. 1977. *Coastal Resource Management: Institutions and Programs.* Ann Arbor: Ann Arbor Science.

Kaufman, W., and O.H. Pilkey, Jr. 1983. *The Beaches Are Moving: The Drowning of America's Shoreline.* Durham: Duke University Press.

Kelley, J.T., et al. 1983. *Living with the Louisiana Shore.* Durham: Duke University Press.

Ketchum, B.H., ed. 1972. *The Water's Edge: Critical Problems of the Coastal Zone.* Cambridge: M.I.T. Press.

Leatherman, S.P. 1979. *Barrier Island Handbook.* National Park Service Special Publication. Amherst: University of Massachusetts, The Environmental Institute.

Leatherman, S.P., ed. 1981. *Overwash Processes.* New York: Van Nostrand Reinhold.

McCormick, L., et al. 1984. *Living with Long Island's South Shore.* Durham: Duke University Press.

Mitchell, J.K. 1974. *Community Response to Coastal Erosion: Individual and Collective Adjustments to Hazard on the Atlantic Shore.* Research Paper no. 156. Chicago: The University of Chicago, Department of Geography.

Morton, R.A., et al. 1983. *Living with the Texas Shore.* Durham: Duke University Press.

Myers, J.C. 1981. *America's Coasts in the 80s: Policies and Issues.* Washington: The Coastal Alliance.

Neal, W.J., et al. 1984. *Living with the South Carolina Shore.* Durham: Duke University Press.

Nordstrom, K.F. 1977. *Coastal Geomorphology of New Jersey.* Vol. I: *Management Techniques and Management Strategies.* Vol. II: *Basis and Background for Management Techniques and Management Strategies.* New Brunswick: Rutgers University, Center for Coastal and Environmental Studies.

Petak, W.J., and A.A. Atkisson. 1982. *Natural Hazard and Public Policy: Anticipating the Unexpected.* New York: Springer-Verlag.

Pilkey, O.H., Jr., W.J. Neal, O.H. Pilkey, Sr., and S.R. Riggs. 1982. *From Currituck to Calabash: Living with North Carolina's Barrier Islands.* Durham: Duke University Press.

Pilkey, O.H., Jr., et al. 1984. *Living with the East Florida Shore.* Durham: Duke University Press.

Pilkey, O.H., Sr., et al. 1983. *Coastal Design: A Guidebook for Builders, Planners and Homeowners.* New York: Van Nostrand Reinhold.

Richardson, D.K. 1976. *The Cost of Environmental Protection: Regulating Housing Development in the Coastal Zone.* New Brunswick: Rutgers University, Center for Urban Policy Research.

Ringold, P.L., and J. Clark. 1980. *The Coastal Almanac.* San Francisco: W.H. Freeman and Company.

Schwartz, M.L., ed. 1982. *The Encyclopedia of Beaches and Coastal Environments.* New York: Van Nostrand Reinhold.

Simon, A.W. 1978. *The Thin Edge: Coast and Man in Crisis.* New York: Harper and Row.

Snead, R.E. 1982. *Coastal Landforms and Surface Features: A Photographic Atlas and Glossary.* New York: Van Nostrand Reinhold.

Sorenson, J.H., and J.K. Mitchell. 1975. *Coastal Erosion Hazards in the United States: A Research Assessment.* Boulder: University of Colorado, Institute of Behavioral Science.
Swift, D.J.P., and H.D. Palmer, eds. 1978. *Coastal Sedimentology.* New York: Van Nostrand Reinhold.
White, G.F., and E.J. Haas. 1975. *Assessment of Research on Natural Hazards.* Cambridge: M.I.T. Press.

JOURNAL ARTICLES

Baker, E.J. 1977. "Public Attitudes toward Hazard Zone Controls." *Journal of the American Institute of Planners* 43(4):401-408.
Baker, E.J., and G. McPhee. 1979. "Geographical Variations in Hurricane Risk and Legislative Response." *Coastal Zone Management Journal* 5(4):263-283.
Burka, P. 1974. "Shoreline Erosion: Implications for Public Rights and Private Ownership." *Coastal Zone Manangement Journal* (2):175-195.
Clark, J., S. McCreary. 1980-81. "Prospects for Coastal Resource Conservation in the 1980's." *Oceanus* 23(4):22-31.
Conservation Foundation. 1977. "The Coasts Are Awash with Disputes." *Conservation Foundation Letter* (March).
Dolan, R., and P. Godfrey. 1973. "Effects of Hurricane Ginger on the Barrier Islands of North Carolina." *Geological Society of America Bulletin* 84(4):1329-1333.
Dolan, R., P. Godfrey, and W.E. Odum. 1973. "Man's Impact on the Barrier Islands of North Carolina." *American Scientist* 61(2):152-166.
Dolan, R., and B. Hayden. 1974. "Adjusting to Nature in Our National Seashores." *National Parks and Conservation Magazine* 48(6):9-14.
_____. 1980-81. "Templates of Change: Storms and Shoreline Hazards." *Oceanus* 23(4):32-37.
Dolan, R., B. Hayden, and J. Heywood. 1978. "Analysis of Coastal Erosion and Storm Surge Hazards." *Coastal Engineering* 2:41-53.
Dolan, R., B. Hayden, and H. Lins. 1980. "Barrier Islands." *American Scientist* 68:16-25.
Godfrey, P.J. 1976. "Barrier Beaches of the East Coast." *Oceanus* 19(5):27-40.
_____. 1978. "Management Guidelines for Parks on Barrier Beaches." *Parks* 2(4):5-10.
Gordon, W.R., Jr. 1984. "The Coastal Barrier Resources Act of 1982: An Assessment of Legislative Intent, Process, and Exemption Alternatives." *Coastal Zone Management Journal* 12(2/3):257-286.
Hildreth, R.G. 1980. "Coastal Natural Hazards Management." *Oregon Law Review* 50(March):201-242.
Inman, D.L., and B.M. Brush. 1973. "The Coastal Challenge." *Science* 181(4094):20-32.
Kuehn, R.R. 1981. "The Shifting Sands of Federal Barrier Islands Policy." *Harvard Environmental Law Review* 5:217-258.
_____. 1984. "The Coastal Barrier Resources Act and the Expenditures Limitation Approach to Natural Resources Conservation: Wave of the Future or Island unto Itself." *Ecology Law Quarterly* 11:583-670.
Leatherman, S.P. 1981. "Barrier Beach Development: A Perspective on the Problem." *Shore and Beach* 49(2):3-9.
McLeish, W.H. 1980. "Our Barrier Islands Are the Key Issue in 1980, the ' Year of the Coast.'" *Smithsonian* 2(6):46-60.
Miller, H.C. 1975. "Coastal Flood Plain Management and the National Flood

Insurance Program: A Case Study of Three Rhode Island Communities." *Environmental Comment* (Nov.):2-14.

_____. 1980-81. "Federal Policies in Barrier Island Development." *Oceanus* 23(4):47-55.

_____. 1981. "The Barrier Islands: A Gamble with Time and Nature." *Environment* (Nov.):6-12; 36-41.

_____. 1982. "Castles in the Sand: Building on Barrier Islands." *Southern Exposure* 10(3):44-48.

Mitchell, J.K. 1976. "Onshore Impacts of Scottish Offshore Oil: Planning Implications for the Middle Atlantic States." *Journal of the American Institute of Planners* 42(4):386-398.

Multiple Authors. 1985. "Symposium: Coastal Zone Management: Planning on the Edge." *Journal of the American Planning Association* 51(3):263-337.

Pilkey, O.H. 1981. "Geologists, Engineers, and a Rising Sea Level." *Northeastern Geology* 3(3/4):150-158.

Platt, R.H. 1978. "Coastal Hazards and National Policy: A Jury-Rig Approach." *Journal of the American Institute of Planners* 4:170-180.

_____. 1985. "Congress and the Coast." *Environment* 27(July/Aug.):12-17; 34-40.

Price, W.A. 1971. "Environmental Impact of Padre Isles Development." *Shore and Beach* 39(2):4-10.

Scott, J.M. 1982. "Coastal Development and Federal Policy: A Case Study." *Urban Land* 41(3):20-27.

Siffin, W.J. 1981. "Bureaucracy, Entrepreneurship, and Natural Resources: Witless Policy and the Barrier Islands." *Cato Journal*(1):294.

Slaughter, T.H. 1973. "Regulatory Aspects Relevant to Coastal Management Problems: Ocean City, Maryland's Coastal Beach." *Shore and Beach* 41(2):5-11.

SELECTED BIBLIOGRAPHY

ARTICLES IN BOOKS—PROCEEDINGS

Baker, E.J., and D.J. Patton. 1974. "Attitudes toward Hurricane Hazard on the Gulf Coast." In *Natural Hazards—Local, National, Global*, ed. G.F. White, pp. 30-36. New York: Oxford University Press.

Beaumann, D.D., and J. H. Sims. 1974. "Human Response to the Hurricane." In *Natural Hazards—Local, National, Global*, ed. G.F. White, pp. 25-30. New York: Oxford University Press.

Brown, P.N., ed. 1979. *The Future of the New Jersey Shore: Problems and Recommended Solutions: Proceedings of the Governor's Conference on the Future of the New Jersey Shore*. New Brunswick, New Jersey.

Dolan, R. 1973. "Barrier Islands: Natural and Controlled." In *Coastal Geomorphology*, ed. D.R. Coates. New York: State University of New York.

Godfrey, P.J. 1977. "Recreational Impact on Shorelines—Research and Management." In *Coastal Recreation Resources in an Urbanized Environment*. Amherst: University of Massachusetts Cooperative Extension Service.

Godfrey, P.J., and M.M. Godfrey. 1973. "Comparison of Ecological and Geomorphic Interactions between Altered and Unaltered Barrier Islands and Systems in North Carolina." In *Coastal Geomorphology*, ed. D.R. Coates. Binghamton: State University of New York Press.

_____. 1975. "Some Estuarine Consequences of Barrier Island Stablization." In *Estuarine Research*, ed. L.E. Cronin, vol. II, pp. 485-516. New York: Academic Press.

Machemehl, J.L. 1978. "Model Building Code for Coastal Hazard Areas." In *Proceedings of Coastal Zone '78*, pp. 1453-1467. New York: American Society of Civil Engineers.

Magoon, O.T., and H. Converse, eds. 1983. *Coastal Zone '84, Proceedings of the Third Symposium on Coastal and Ocean Management, San Diego, June 1-4, 1983*, 3 vols. New York: American Society of Civil Engineers.

Miller, H.C. 1976. "Barrier Islands, Barrier Beaches, and the National Flood Insurance Program: Some Problems and a Rationale for Special Attention." In *Barrier Islands and Beaches: Technical Proceedings of the 1976 Barrier Island Workshop*. Annapolis, Maryland.

Mitchell, J.K. 1972. "Global Summary of Human Response to Coastal Erosion." In *22nd International Geographical Congress*, paper no. 9. Boulder: University of Colorado, Institute of Behavioral Sciences.

_____. 1982. "Coastal Zone Management: A Comparative Analysis of National Programs." In *Ocean Yearbook 3*, ed. E.M. Borgese and N.S. Ginsburg, pp. 258-319. Chicago: University of Chicago Press.

Rountree, R.A. 1974. "Coastal Erosion: The Meaning of a Natural Hazard in the Cultural and Ecological Context." In *Natural Hazards—Local, National, Global*, ed. G.F. White, pp. 70-79. New York: Oxford University Press.

White, A.U. 1974. "Global Summary of Human Response to Natural Hazards: Tropical Cyclones." In *Natural Hazards—Local, National, Global*, ed. G.F. White, pp. 225-264. New York: Oxford University Press.

SELECTED BIBLIOGRAPHY

REPORTS AND MONOGRAPHS

Clark, J.F. 1974. *Coastal Ecosystems: Ecological Considerations for Management of the Coastal Zone.* Washington: The Conservation Foundation.

──────────────. 1976. *The Sanibel Report: Formulation of a Comprehensive Plan Based on Natural Systems.* Washington: The Conservation Foundation.

Conservation Foundation. 1977. *Physical Management of Coastal Floodplains: Guidelines for Hazard and Ecosystems Management.* Washington: The Conservation Foundation.

──────────────. 1980. *Coastal Environmental Management: Guidelines for Conservation of Resources and Protection against Storm Hazards.* Washington: U.S. Government Printing Office.

Federal Insurance Administration. 1977. *Proceedings of the National Conference on Coastal Erosion, July 6-8, 1977.* Washington: Federal Insurance Administration.

Kusler, J.A. 1982. *Regulation of Flood Hazard Areas to Reduce Flood Losses, Vol. 3.* Special Publication 2. Boulder: University of Colorado Natural Hazards Research and Applications Information Center.

Leatherman, S.P., and P.J. Godfrey. 1979. *The Impact of Off-Road Vehicles on Coastal Ecosystems in Cape Cod National Seashore: An Overview.* Amherst: National Park Service Cooperative Research Unit, University of Massachusetts.

Lins, H.F. 1980. *Patterns and Trends of Land-Use and Land Cover on Atlantic and Gulf Coast Barrier Islands.* USGS Prof. Paper no. 1156. Washington: U.S. Government Printing Office.

Massachusetts Office of Coastal Zone Management. 1985. *The Way to the Sea.* Boston: Mass. OCZM.

National Research Council. 1984. *Hurricane Alicia: Galveston and Houston, Texas, August 17-18, 1983.* Washington: National Academy Press.

New England River Basins Commission. 1979. *Dealing with Coastal Hazards: Implementing the Regional Policy Statement on Flood Plain Management.* Boston: NERBC.

New Jersey Beach Access Study Commission. 1977. *Public Access to the Oceanfront Beaches: A Report to the Governor and Legislature of New Jersey.* Trenton: The Commission.

New Jersey Department of Environmental Protection, Division of Coastal Resources. 1982. *Coastal Resources and Development Policies.* Trenton: DEP.

Platt, R.H., and G.M. McMullen. 1980. *Post-Flood Recovery and Hazard Mitigation: Lessons from the Massachusetts Coast, February 1978.* Report no. 115. Amherst: University of Massachusetts, Water Resources Research Center.

Reynolds, M.O., M. Kaufman, and T. Cohen. *Living by the Sea: A Massachusetts Citizen's Handbook for Coastal Zone Management Planning.* Boston: Massachusetts Office of Coastal Zone Management.

Rogers, J., and F. Golden. 1979. *Coastal Development.* Trenton: New Jersey

Department of Environmental Protection.

Rubin, C.B. 1981. *Long-Term Recovery from Natural Disasters: A Comparative Analysis of Six Local Experiences.* Washington: Academy for Contemporary Problems.

U.S. Army Corps of Engineers. 1971. *Report on the National Shoreline Study.* Washington: The Corps.

U.S. Department of Commerce, Office of Coastal Zone Management. 1976. *Natural Hazard Management in Coastal Areas.* Washington: OCZM.

_____. 1976. *Who's Minding the Shore? A Citizen's Guide to Coastal Management.* Washington: OCZM.

_____. 1979. *The First Five Years of Coastal Zone Management. An Initial Assessment.* Washington: OCZM.

U.S. Department of the Interior. 1980. *Alternative Policies for Protecting Barrier Islands along the Atlantic and Gulf Coasts of the United States and Draft Environmental Impact Statement.* Washington: DOI.

_____. 1982. *Undeveloped Coastal Barriers: Report to Congress.* Washington: U.S. Government Printing Office.

_____. 1985. *Coastal Barrier Resources System: Draft Report to Congress.* Washington: U.S. Government Printing Office.

U.S. General Accounting Office. 1982. *National Flood Insurance: Marginal Impact on Floodplain Development.* (CED-82-105). Washington: GAO.

THE UNIVERSITY OF CHICAGO
DEPARTMENT OF GEOGRAPHY
RESEARCH PAPERS (Lithographed, 6 × 9 inches)

LIST OF TITLES IN PRINT

48. BOXER, BARUCH. *Israeli Shipping and Foreign Trade.* 1957. 162 p.
56. MURPHY, FRANCIS C. *Regulating Flood-Plain Development.* 1958. 216 pp.
62. GINSBURG, NORTON, editor. *Essays on Geography and Economic Development.* 1960. 173 p.
71. GILBERT, EDMUND WILLIAM. *The University Town in England and West Germany.* 1961. 79 p.
72. BOXER, BARUCH. *Ocean Shipping in the Evolution of Hong Kong.* 1961. 108 p.
91. HILL, A. DAVID. *The Changing Landscape of a Mexican Municipio, Villa Las Rosas, Chiapas.* 1964. 121 p.
97. BOWDEN, LEONARD W. *Diffusion of the Decision To Irrigate: Simulation of the Spread of a New Resource Management Practice in the Colorado Northern High Plans.* 1965. 146 pp.
98. KATES, ROBERT W. *Industrial Flood Losses: Damage Estimation in the Lehigh Valley.* 1965. 76 pp.
101. RAY, D. MICHAEL. *Market Potential and Economic Shadow: A Quantitative Analysis of Industrial Location in Southern Ontario.* 1965. 164 p.
102. AHMAD, QAZI. *Indian Cities: Characteristics and Correlates.* 1965. 184 p.
103. BARNUM, H. GARDINER. *Market Centers and Hinterlands in Baden-Württemberg.* 1966. 172 p.
105. SEWELL, W. R. DERRICK, et al. *Human Dimensions of Weather Modification.* 1966. 423 p.
107. SOLZMAN, DAVID M. *Waterway Industrial Sites: A Chicago Case Study.* 1967. 138 p.
108. KASPERSON, ROGER E. *The Dodecanese: Diversity and Unity in Island Politics.* 1967. 184 p.
109. LOWENTHAL, DAVID, editor, *Environmental Perception and Behavior.* 1967. 88 p.
112. BOURNE, LARRY S. *Private Redevelopment of the Central City, Spatial Processes of Structural Change in the City of Toronto.* 1967. 199 p.
113. BRUSH, JOHN E., and GAUTHIER, HOWARD L., JR., *Service Centers and Consumer Trips: Studies on the Philadelphia Metropolitan Fringe.* 1968. 182 p.
114. CLARKSON, JAMES D., *The Cultural Ecology of a Chinese Village: Cameron Highlands, Malaysia.* 1968. 174 p.
115. BURTON, IAN, KATES, ROBERT W., and SNEAD, RODMAN E. *The Human Ecology of Coastal Flood Hazard in Megalopolis.* 1968. 196 p.
117. WONG, SHUE TUCK, *Perception of Choice and Factors Affecting Industrial Water Supply Decisions in Northeastern Illinois.* 1968. 93 p.
118. JOHNSON, DOUGLAS L. *The Nature of Nomadism: A Comparative Study of Pastoral Migrations in Southwestern Asia and Northern Africa.* 1969. 200 p.
119. DIENES, LESLIE. *Locational Factors and Locational Developments in the Soviet Chemical Industry.* 1969. 262 p.
120. MIHELIČ, DUŠAN. *The Political Element in the Port Geography of Trieste.* 1969. 104 p.
121. BAUMANN, DUANE D. *The Recreational Use of Domestic Water Supply Reservoirs: Perception and Choice.* 1969. 125 p.
122. LIND, AULIS O. *Coastal Landforms of Cat Island, Bahamas: A Study of Holocene Accretionary Topography and Sea-Level Change.* 1969. 156 p.
123. WHITNEY, JOSEPH B. R. *China: Area, Administration and Nation Building.* 1970. 198 p.
124. EARICKSON, ROBERT. *The Spatial Behavior of Hospital Patients: A Behavioral Approach to Spatial Interaction in Metropolitan Chicago.* 1970. 138 p.
125. DAY, JOHN CHADWICK. *Managing the Lower Rio Grande: An Experience in International River Development.* 1970. 274 p.
126. MacIVER, IAN. *Urban Water Supply Alternatives: Perception and Choice in the Grand Basin Ontario.* 1970. 178 p.
127. GOHEEN, PETER G. *Victorian Toronto, 1850 to 1900: Pattern and Process of Growth.* 1970. 278 p.
128. GOOD, CHARLES M. *Rural Markets and Trade in East Africa.* 1970. 252 p.
129. MEYER, DAVID R. *Spatial Variation of Black Urban Households.* 1970. 127 p.
130. GLADFELTER, BRUCE G. *Meseta and Campiña Landforms in Central Spain: A Geomorphology of the Alto Henares Basin.* 1971. 204 p.
131. NEILS, ELAINE M. *Reservation to City: Indian Migration and Federal Relocation.* 1971. 198 p.
132. MOLINE, NORMAN T. *Mobility and the Small Town, 1900–1930.* 1971. 169 p.

133. SCHWIND, PAUL J. *Migration and Regional Development in the United States.* 1971. 170 p.
134. PYLE, GERALD F. *Heart Disease, Cancer and Stroke in Chicago: A Geographical Analysis with Facilities, Plans for 1980.* 1971. 292 p.
135. JOHNSON, JAMES F. *Renovated Waste Water: An Alternative Source of Municipal Water Supply in the United States.* 1971. 155 p.
136. BUTZER, KARL W. *Recent History of an Ethiopian Delta: The Omo River and the Level of Lake Rudolf.* 1971. 184 p.
139. MCMANIS, DOUGLAS R. *European Impressions of the New England Coast, 1497–1620.* 1972. 147 p.
140. COHEN, YEHOSHUA S. *Diffusion of an Innovation in an Urban System: The Spread of Planned Regional Shopping Centers in the United States, 1949–1968,* 1972. 136 p.
141. MITCHELL, NORA. *The Indian Hill-Station: Kodaikanal.* 1972. 199 p.
142. PLATT, RUTHERFORD H. *The Open Space Decision Process: Spatial Allocation of Costs and Benefits.* 1972. 189 p.
143. GOLANT, STEPHEN M. *The Residential Location and Spatial Behavior of the Elderly: A Canadian Example.* 1972. 226 p.
144. PANNELL, CLIFTON W. *T'ai-chung, T'ai-wan: Structure and Function.* 1973. 200 p.
145. LANKFORD, PHILIP M. *Regional Incomes in the United States, 1929–1967: Level, Distribution, Stability, and Growth.* 1972. 137 p.
146. FREEMAN, DONALD B. *International Trade, Migration, and Capital Flows: A Quantitative Analysis of Spatial Economic Interaction.* 1973. 201 p.
147. MYERS, SARAH K. *Language Shift Among Migrants to Lima, Peru.* 1973. 203 p.
148. JOHNSON, DOUGLAS L. *Jabal al-Akhdar, Cyrenaica: An Historical Geography of Settlement and Livelihood.* 1973. 240 p.
149. YEUNG, YUE-MAN. *National Development Policy and Urban Transformation in Singapore: A Study of Public Housing and the Marketing System.* 1973. 204 p.
150. HALL, FRED L. *Location Criteria for High Schools: Student Transportation and Racial Integration.* 1973. 156 p.
151. ROSENBERG, TERRY J. *Residence, Employment, and Mobility of Puerto Ricans in New York City.* 1974. 230 p.
152. MIKESELL, MARVIN W., editor. *Geographers Abroad: Essays on the Problems and Prospects of Research in Foreign Areas.* 1973. 296 p.
153. OSBORN, JAMES F. *Area, Development Policy, and the Middle City in Malaysia.* 1974. 291 p.
154. WACHT, WALTER F. *The Domestic Air Transportation Network of the United States.* 1974. 98 p.
155. BERRY, BRIAN J. L., et al. *Land Use, Urban Form and Environmental Quality.* 1974. 440 p.
156. MITCHELL, JAMES K. *Community Response to Coastal Erosion: Individual and Collective Adjustments to Hazard on the Atlantic Shore.* 1974. 209 p.
157. COOK, GILLIAN P. *Spatial Dynamics of Business Growth in the Witwatersrand.* 1975. 144 p.
159. PYLE, GERALD F. et al. *The Spatial Dynamics of Crime.* 1974. 221 p.
160. MEYER, JUDITH W. *Diffusion of an American Montessori Education.* 1975. 97 p.
161. SCHMID, JAMES A. *Urban Vegetation: A Review and Chicago Case Study.* 1975. 266 p.
162. LAMB, RICHARD F. *Metropolitan Impacts on Rural America.* 1975. 196 p.
163. FEDOR, THOMAS STANLEY. *Patterns of Urban Growth in the Russian Empire during the Nineteenth Century.* 1975. 245 p.
164. HARRIS, CHAUNCY D. *Guide to Geographical Bibliographies and Reference Works in Russian or on the Soviet Union.* 1975. 478 p.
165. JONES, DONALD W. *Migration and Urban Unemployment in Dualistic Economic Development.* 1975. 174 p.
166. BEDNARZ, ROBERT S. *The Effect of Air Pollution on Property Value in Chicago.* 1975. 111 p.
167. HANNEMANN, MANFRED. *The Diffusion of the Reformation in Southwestern Germany, 1518–1534.* 1975. 248 p.
168. SUBLETT, MICHAEL D. *Farmers on the Road. Interfarm Migration and the Farming of Noncontiguous Lands in Three Midwestern Townships. 1939–1969.* 1975. 228 pp.
169. STETZER, DONALD FOSTER. *Special Districts in Cook County: Toward a Geography of Local Government.* 1975. 189 pp.
171. SPODEK, HOWARD. *Urban-Rural Integration in Regional Development: A Case Study of Saurashtra, India—1800–1960.* 1976. 156 pp.
172. COHEN, YEHOSHUA S. and BERRY, BRIAN J. L. *Spatial Components of Manufacturing Change.* 1975. 272 pp.
173. HAYES, CHARLES R. *The Dispersed City: The Case of Piedmont, North Carolina.* 1976. 169 pp.

174. CARGO, DOUGLAS B. *Solid Wastes: Factors Influencing Generation Rates.* 1977. 112 pp.
175. GILLARD, QUENTIN. *Incomes and Accessibility. Metropolitan Labor Force Participation, Commuting, and Income Differentials in the United States, 1960–1970.* 1977. 140 pp.
176. MORGAN, DAVID J. *Patterns of Population Distribution: A Residential Preference Model and Its Dynamic.* 1978. 216 pp.
177. STOKES, HOUSTON H.; JONES, DONALD W. and NEUBURGER, HUGH M. *Unemployment and Adjustment in the Labor Market: A Comparison between the Regional and National Responses.* 1975. 135 pp.
179. HARRIS, CHAUNCY D. *Bibliography of Geography. Part I. Introduction to General Aids.* 1976. 288 pp.
180. CARR, CLAUDIA J. *Pastoralism in Crisis. The Dasanetch and their Ethiopian Lands.* 1977. 339 pp.
181. GOODWIN, GARY C. *Cherokees in Transition: A Study of Changing Culture and Environment Prior to 1775.* 1977. 221 pp.
182. KNIGHT, DAVID B. *A Capital for Canada: Conflict and Compromise in the Nineteenth Century.* 1977. 359 pp.
183. HAIGH, MARTIN J. *The Evolution of Slopes on Artificial Landforms: Blaenavon, Gwent.* 1978. 311 pp.
184. FINK, L. DEE. *Listening to the Learner. An Exploratory Study of Personal Meaning in College Geography Courses.* 1977. 200 pp.
185. HELGREN, DAVID M. *Rivers of Diamonds: An Alluvial History of the Lower Vaal Basin.* 1979. 399 pp.
186. BUTZER, KARL W., editor. *Dimensions of Human Geography: Essays on Some Familiar and Neglected Themes.* 1978. 201 pp.
187. MITSUHASHI, SETSUKO. *Japanese Commodity Flows.* 1978. 185 pp.
188. CARIS, SUSAN L. *Community Attitudes toward Pollution.* 1978. 226 pp.
189. REES, PHILIP M. *Residential Patterns in American Cities, 1960.* 1979. 424 pp.
190. KANNE, EDWARD A. *Fresh Food for Nicosia.* 1979. 116 pp.
192. KIRCHNER, JOHN A. *Sugar and Seasonal Labor Migration: The Case of Tucumán, Argentina.* 1980. 158 pp.
193. HARRIS, CHAUNCY D. and FELLMANN, JEROME D. *International List of Geographical Serials, Third Edition, 1980.* 1980. 457 p.
194. HARRIS, CHAUNCY D. *Annotated World List of Selected Current Geographical Serials, Fourth, Edition.* 1980. 1980. 165 p.
195. LEUNG, CHI-KEUNG. *China: Railway Patterns and National Goals.* 1980. 235 p.
196. LEUNG, CHI-KEUNG and GINSBURG, NORTON S., eds. *China: Urbanization and National Development.* 1980. 280 p.
197. DAICHES, SOL. *People in Distress: A Geographical Perspective on Psychological Well-being.* 1981. 199 p.
198. JOHNSON, JOSEPH T. *Location and Trade Theory: Industrial Location, Comparative Advantage, and the Geographic Pattern of Production in the United States.* 1981. 107 p.
199-200. STEVENSON, ARTHUR J. *The New York-Newark Air Freight System.* 1982. 440 p.
201. LICATE, JACK A. *Creation of a Mexican Landscape: Territorial Organization and Settlement in the Eastern Puebla Basin, 1520–1605.* 1981. 143 p.
202. RUDZITIS, GUNDARS. *Residential Location Determinants of the Older Population.* 1982. 117 p.
203. LIANG, ERNEST P. *China: Railways and Agricultural Development, 1875–1935.* 1982. 186 p.
204. DAHMANN, DONALD C. *Locals and Cosmopolitans: Patterns of Spatial Mobility during the Transition from Youth to Early Adulthood.* 1982. 146 p.
205. FOOTE, KENNETH E. *Color in Public Spaces: Toward a Communication-Based Theory of the Urban Built Environment.* 1983. 153 p.
206. HARRIS, CHAUNCY D. *Bibliography of Geography. Part II: Regional. Vol. 1. The United States of America.* 1984. 178 p.
207-208. WHEATLEY, PAUL. *Nāgara and Commandery: Origins of the Southeast Asian Urban Traditions.* 1983. 473 p.
209. SAARINEN, THOMAS F.; SEAMON, DAVID; and SELL, JAMES L., eds. *Environmental Perception and Behavior: An Inventory and Prospect.* 1984. 263 p.
210. WESCOAT, JAMES L., JR. *Integrated Water Development: Water Use and Conservation Practice in Western Colorado.* 1984. 239 p.
211. DEMKO, GEORGE J., and FUCHS, ROLAND J., eds. *Geographical Studies on the Soviet Union: Essays in Honor of Chauncy D. Harris.* 1984. 294 p.

212. HOLMES, ROLAND C. *Irrigation in Southern Peru: The Chili Basin.* 1986. 191 p.
213. EDMONDS, RICHARD L. *Northern Frontiers of Qing China and Tokugawa Japan: A Comparative Study of Frontier Policy.* 1985. 155 p.
214. FREEMAN, DONALD B., and NORCLIFFE, GLEN B. *Rural Enterprise in Kenya: Development and Spatial Organization of the Nonfarm Sector.* 1985. 180 p.
215. COHEN, YEHOSHUA S., and SHINAR, AMNON. *Neighborhoods and Friendship Networks: A Study of Three Residential Neighborhoods in Jerusalem.* 1985. 129 p.
217-218. CONZEN, MICHAEL P., ED. *World Patterns of Modern Urban Change: Essays in Honor of Chauncy D. Harris.* 1986.
219. KOMOGUCHI, YOSHIMI. *Agricultural Systems in the Tamil Nadu: A Case Study of Peruvalanallur Village.* 1986. 171 p.
220. GINSBURG, NORTON; OBORN, JAMES; and BLANK, GRANT. *Geographic Perspectives on the Wealth of Nations.* 1986. 131 p.
221. BAYLSON, JOSHUA C. *Terrritorial Allocation by Imperial Rivalry: The Human Legacy in the Near East.* 1987. 129 p.
224. PLATT, RUTHERFORD H.; PELCZARSKI, SHEILA G.; and BURBANK, BARBARA K. R. EDS. *Cities on the Beach: Management Issues of Developed Coastal Barriers.* 1987. 300 p.